I0048937

WORKING IN
SLIPPERS

WORKING IN
SLIPPERS

Virtual Companies and
the New Normal

KEN TAYLOR

Published by Ripples Media

www.ripples.media
Atlanta, GA

Copyright © 2025 by Ken Taylor

All rights reserved. No part of this book may be repro-
duced or used in any manner without written permission of the copyright owner except for the use of
quotations in a book review. For more information:
publishing@ripples.media

First printing 2025

Cover design & book interior by Carolyn Asman

ISBN 979-8-9997153-1-9 Paperback

ISBN 979-8-9997153-2-6 Hardback

ISBN 979-8-9997153-0-2 E-book

Library of Congress Control Number: 2025918180

DEDICATION

This book is dedicated to my father
and grandfather, who both wrote
many books.

And to my son, who has books in his
head that I can't wait to read.

FOREWORD

BY ADAM WALKER

Like many, I started my company in a basement out of necessity. With two small kids and another on the way, I had been laid off from a job I didn't like, and it was time to start something of my own. As the company grew, I found a partner to merge with, but he lived on the other side of the city. Instead of finding a physical workspace, we relied on technology, using Skype to collaborate and grow the business. As we added employees scattered across the metro area, we continued to use Skype and chat tools to work together. In the early days, we were on Skype all day, collaborating and getting things done.

Believe it or not, this was back in 2008.

During those years, working remotely was a lifestyle choice and a privilege. We were scrappy, which is a charitable way to say that we were barely making it. So, every dollar that wasn't used to pay a team member, needed to go into our pocket. This meant that a physical office space was never really an option (or even on our radar). We just didn't see the point.

Over time, we realized that working from home was a competitive advantage. We were more focused and operated leaner. Most importantly, I could be present for my kids, hugging them as they ran past my desk, getting them on and off the school bus, and being a part of their lives during the workday, even for a moment.

This was more valuable to me than any profit, culture, or growth metric. Yet, working remotely also helped us grow the company faster and increase profits.

Many believe you can't build a good company culture remotely, but that's never been true for me. I've led multiple remote teams over the years, building strong cultures and relationships with people I've never met in person. It's about using the tools available, connecting meaningfully, and enjoying life. It's foolish to think that proximity alone creates culture. That's like saying you don't love your grandma if you don't see her once a week. Building culture, in person or remote, takes thoughtfulness and intentionality; that's it. And that's exactly what this book is here to help you do.

I started working from home in 2005, and built my first remote company starting in 2008. Long before COVID, I didn't understand why small companies felt the need for office space. It seemed like an unnecessary expense, with time wasted on commuting and office chitchat. For over 20 years, I've grown and sold virtual companies, led remote teams, and done meaningful work, all without a physical office.

Through this book, Ken Taylor validates what I learned through trial and error over two decades. His systematic approach to remote work management, particularly around hiring practices and performance measurement, addresses the gaps I wish I had understood early in my journey. And, even now, with 20 years of remote work behind me, his thinking and strategies are still helpful, there is always more to learn.

The advantages Ken outlines in this book aren't just theoretical; I lived them. But he goes deeper, showing how remote work creates a competitive advantage that traditional "office-bound" companies struggle to cross. I've found the advantage isn't just for the companies, but also for the team members who get more flexibility and autonomy, leading to a more satisfying career and life.

Like the rest of the world, during the COVID-19 pandemic, we took our nonprofit event, 48in48, fully remote as well. Previously, we'd traveled the world, hosting in-person events to build 48 free nonprofit websites in 48 hours in cities like Boston, Miami, and London. When COVID hit, our executive director pivoted the event to a virtual format. Since then, we've worked with thousands of volunteers across more than 20 time zones, building hundreds of websites for nonprofits across the world. This experience proved that even short-term, new virtual teams can be productive, create value, build relationships, and foster culture—all without being in the same room. It has enhanced our organization's reach and deepened its impact.

In this book, Ken also addresses culture head-on. Culture is about shared values, clear communication, and mutual respect. Remote-work culture is also largely about trust and giving team members the autonomy and freedom they need not only to succeed, but to grow. Ken illustrates how these actually flourish when you hire the right people and give them the tools and trust to succeed.

Remote work forced me to become a better leader faster. Without the crutch of physical presence, you have to be clearer about expectations, more intentional about communication, and more focused on results. This book provides the roadmap I wish I'd had.

Ken also illuminates how remote companies aren't just office companies without offices, they're an entirely different species of organization, one that's often more resilient, more agile, and more human than their traditional counterparts.

If you're ready to lead with intention, build a thriving culture, and unlock the full potential of remote work, you're in the right place. This book is your guide to building something truly extraordinary—no office required.

—Adam Walker

Husband, Father of 5, Wearer of Fedoras, Helper of Nonprofits, Co-Founder of Sideways8 Interactive, Former CMO of TechBridge, Co-Founder and Executive Director of 48in48, and Founder and Head of Marketing and Operations for Edgewise Media

CONTENTS

Your Secret Weapon

Our Friends Abroad

Less Time with Lawyers

What, Never? Well, Hardly Ever

Roads Not Taken

INTRODUCTION

Do you remember when you first realized that COVID-19 was going to turn the world upside down? For me, it was when the President gave a speech announcing that all travel from Europe would cease. Say *what?* Even during World War II, you could travel to and from Europe if you had enough money.

In the following weeks, sure enough, everything changed. "Two weeks to bend the curve" turned into months and years. Children went home from school like normal that day but never returned to their classroom. Thousands of office workers did the same, abruptly having to figure out how to work, manage, meet, and make money in an unfamiliar environment. The newly homebound realized that toilet paper demand would explode because everybody would be spending almost all their time there, and stores were stripped bare.

While my employer, OwnerRez, suffered COVID heartburn like everyone else, the concept of working from home didn't present a problem. We'd already been doing that! In fact, OwnerRez has never occupied a physical office, and likely never will. We built our structures around the assumption that everyone would work from their own home, wherever that might be.

Without wanting to jinx the future, I can say that approach has been successful. From three people, our organization grew by more than an order of magnitude in five years without any outside investment, bank loans, or other financial burdens. All the growth has been purely organic, funded entirely by earnings generated through hard work for satisfied clients.

While this book can't cover everything that contributed to that success, it concentrates on one key element: how a fully-remote culture can thrive and contribute to overall corporate success in a sustainable way.

CHAPTER 1

A WORLD OF CHANGE
The Pendulum Swings Back?

Build a successful, growing company without a physical office space where staff see peers, subordinates, and bosses every day? Most titans of American business think this is foolishness. Jamie Dimon of banking giant JP Morgan Chase is well known for his hatred of remote work[1], dismissing it as something useful "to help women stay home a little." Billionaire entrepreneur Elon Musk has announced[2], "Remote work is no longer acceptable." Even the federal government, not noted for being particularly concerned with employee productivity, is under pressure to drag employees back[3] into the office.

[1] Cheng, Michelle. "Jamie Dimon wasn't kidding about his hatred of remote work." Yahoo Tech! https://tech.yahoo.com/business/articles/jamie-dimon-wasn-t-kidding-101327661.html Accessed Apr 22, 2025.
[2] Hetzner, Christiaan. "'Pretend to work somewhere else': Elon Musk's leaked email ends remote work privileges for Tesla staff." Fortune Magazine. https://fortune.com/2022/06/01/tesla-elon-musk-work-from-home-remote-office-hybrid/ Accessed Apr 22, 2025.
[3] Friedman, Drew. "White House tells agencies to strike a balance between telework, in-office work." Federal News Network. https://federalnewsnetwork.com/workforce/2023/04/white-house-tells-agencies-to-strike-a-balance-between-telework-in-office-work/ Accessed Apr 22, 2025.

Do Mr. Dimon and Mr. Musk know something I don't? Of course: they know their own employees and businesses far better than I ever will. It's entirely possible that those businesses and those employees cannot maintain a healthy, productive corporate culture based on working from home, even though they quickly adopted some of the same tools and technologies that made OwnerRez and companies like it fully remote.

This isn't new. IBM invented the PC but couldn't reinvent their business to center around a technology that had to be developed, marketed, and sold in a wildly different fashion than the mainframe computing that was their core business at the time. Startup companies like Apple and Microsoft, not burdened by old assumptions and legacy practices, instead built the platforms and fortunes of the next generation.

Similarly, I can't say whether it's possible to convert a legacy, traditional business into a virtual company. Mr. Dimon and Mr. Musk believe it is not, and I'm inclined to take their views seriously. But for a new business or an existing firm without such strongly-entrenched cultures and unencumbered with traditional office structures, that's a different matter. I know it can be done. I've done it.

—— From The Beginning ——

Our company didn't start out with the conscious decision to always be a fully-remote business. It was just the natural result of a business run by two founders who lived on opposite sides of the country and, thanks to a background of independent software development, were accustomed to working from their own home offices already. When I joined as the third member, I briefly co-located with one of the founders at their personal home for knowledge transfer, but after a few weeks, it didn't seem worth the drive when we could talk on Google Meet whenever we needed.

As we started to grow, most of the initial new hires, by happenstance, weren't close to any existing leader's home, so they naturally wound up working out of their own homes as well. Any growing business must establish policies and procedures that suit its needs. With OwnerRez, that happened to be remote. Then came COVID, and we realized that establishing a conventional office wouldn't be practical anytime soon, even if we wanted to. In this way, the remote work culture was "baked in" to what OwnerRez became.

Aside from COVID, another motivating factor to stick with a remote model was the challenge of finding the best possible hires. Outside of famous tech hubs like Silicon Valley, a surprising number of people are reluctant to join a small startup in the first place. To establish an effective office, the space has to be located within reasonable commuting range of an adequate supply of workers.

The trouble is, most U.S. cities developed along a radial model, with the idea of everyone commuting into the city center, where office space is ludicrously expensive. In the suburbs, office space is much cheaper, but public transportation is rare, and the roads don't make commutes easy. This further reduces the available choices of candidates.

One well-known source of excellent prospects is through referrals from existing employees. At one time, that would naturally have meant that most candidates lived nearby. Nowadays, though, most people's friend groups span a far broader area. If nothing else, their college buddies could be from anywhere.

This proved to be another significant driver of our fully-remote practices: the best people we were able to find generally didn't live anywhere close. Is it better to hire the most appealing candidates and deal with the novelty of fully-remote work or is the company best served by settling for those who live close by?

This calculation was complicated for our organization because, at the time, the founders lived in cities that are notably expensive. Employees also living in that area would naturally expect the high cost of living to be reflected in higher paychecks.

On the other hand, OwnerRez is in the business of providing operating software for vacation rental properties. Some of those properties are found in every major city, of course, but far more are located in vacation areas near mountains, lakes, and forests, where the cost of living is considerably less. Hiring employees who already have experience with your products is advantageous, and our prospects typically lived far from major cities or any particular office.

As we grew organically, the best candidates cropped up in unexpected places. Occasionally, we'd find a cluster – there seem to be quite a few excellent and experienced customer success representatives throughout the state of Florida and the metropolitan area of Atlanta, for example – but never enough to warrant concentrating everything in an expensive physical office.

With each new hire that didn't live near anyone else, the permanent need to support remote work became more obvious. What alternatives were there? We could have paid for everyone to move to one place, but not everyone was willing to relocate, including the founders themselves, who were already living far apart. We could have laid off those who didn't want to move, but that was never a serious option. Hiring good people is hard enough, so it would have made no sense to eliminate those who had already proven their worth just because they lived in the "wrong" place.

As time went on, the fully-remote model matured, first by happenstance, then increasingly by design. The advantages became clearer, contributing to the success of the company as well as to the experience of each employee.

The key to this innovation is recognizing that a remote company is simply a new manifestation of what every successful company already knows: its people are indeed its most important resource.

CHAPTER 2

SETTING THE OLD ASIDE

Executive Leadership's Greatest Challenge

"Nine Out Of 10 Companies Will Require Employees Return to the Office!" screams *Forbes*[4] magazine. Every day brings a new headline of a CEO who's decided that although they had to tolerate work-from-home at the height of the pandemic, that's over now, and it's time to return to business as "usual." Could it be that remote work was just a temporary phenomenon, one that will soon be just a fading memory for those who lived through it? In some companies, perhaps it was. But remote work is a permanent addition to business in the long term, and here's why.

---

[4] Hyken, Shep. "Nine Out Of 10 Companies Will Require Employees To Return To The Office." *Forbes* magazine. https://www.forbes.com/sites/shephyken/2023/09/24/nine-out-of-10-companies-will-require-employees-to-return-to-the-office/ Accessed Apr 22, 2025.

—— Keep On Keeping On ——

Consider senior management at a large company, just under the C-suite. VPs are often tasked with building policies that reinforce the CEO's leadership and vision. While their power is limited, it is quite broad.

How did they rise to that almost-exalted position? In most cases, they navigated years of corporate ladder climbing, avoiding mistakes, making job moves at the magic moment, and seizing ripe professional opportunities as they arose. And for nearly all of their career, that work was done *in an office*, because that's largely how business has been done for living memory.

A corporate vice president has already proven successful in an office environment. In a no-office environment, they might also thrive – or perhaps not. Should they risk their established career by completely changing their familiar operational environment? COVID gave them no choice. Offices were closed by government fiat, requiring quick fixes for novel situations. Now that those edicts are history, companies seek to bring their operations back to what they know so well.

We've seen similar patterns in other technological revolutions. Even years after email became commonplace, many senior executives still relied on assistants to print their emails for review – a practice that seems almost quaint now.

For a large company to completely change how it does business is *extremely difficult*. Most resist doing so as hard as they can, and when they're finally forced to, they generally fail. The same is true of tenured executives who've been operating a certain way for a long time.

A few decades ago, Walmart was one of the world's most valuable companies. Today, Amazon has eclipsed its dominance.

Walmart is not small. They still make large profits, and they've dumped huge sums and tremendous effort into launching their own online sales sites to compete with Amazon. Yet their online sales are a fraction of those of Amazon. Despite having plenty of money and a growing awareness of Amazon's long-term threat, Walmart is unlikely to successfully transform itself into a fully online, or even an online-first, company, in a way that would allow them to be a true threat to Amazon rather than a far-smaller also-ran.

Does this mean that Walmart is at risk of bankruptcy? No, they probably have many decades of profitability remaining, and then potentially several more decades of decline. A similar fate happened to Sears and K-Mart over an entire lifetime, caused largely by upstart Walmart's success. It's also possible that they'll survive indefinitely, as surely not all sales will be exclusively online! JC Penney and Macy's still survive, unlike Sears, K-Mart, and many more. But they do not command the headlines and customer awareness the way Sears once did, and are hardly high-growth companies the way WalMart was thirty years ago.

It takes a really long time to kill off a giant corporation. Most likely everyone now employed by Walmart would be retired before serious struggles are plainly visible. For people who've already invested significant portions of their career in Walmart as it's been, fully reinventing the entire company's operation would be a dubious company- and career-gambling high-risk strategy.

Hidden Costs of Short-Sighted Choices

The same is true with remote work. Companies whose executives perceive risks in letting their employees work from home *can* require them to return to the office, but they may be underestimating both the short-term and long-term costs of this decision.

Who Quits First?

Consider a traditional white-collar office-based company that established a work-from-home policy at the height of COVID. Some employees have been champing at the bit to return to the office. Some chose to move halfway across the country or even the world and are "living their best lives" in the mountains or on the beach with a high-speed Internet connection.

When the boss announces everyone has to be back in the office, what happens to the employees now living a thousand miles away who've been doing good remote work for years? They face a difficult decision that can only harm the company. They may reluctantly sell their dream home and move back to a standard suburban tract house. Are they going to feel happy at their desk each day? Will they likely be inspired to do their best work? Or will they experience bitterness and resentment?

What would lead them to upend their lives so dramatically? Simple: they don't believe that they can find comparable employment. In effect, they have the best-paying job they think they can find.

Now, some people are naturally conservative and risk-averse, unwilling to let go of an apparent good thing, even at the expense of their way of life. Often, though, the employees who don't think they could find a comparable job are *right* because they really aren't that good. So those are the ones who will stay!

How about those who are confident in their own competence and capabilities? They will likely seek remote employment elsewhere with some other company that preserves their remote way of life. We already saw some of this with the so-called "Great Resignation."

Do you see the difference? The *best* employees will be the *least* likely to accept a demand to return to the office. The least productive will be the most likely to accept it because they feel they have no choice. A return-to-office demand will most likely result in diminished employee quality. How is this beneficial?

What's more, this effect feeds into one of the greatest benefits of remote work that we discussed earlier: the ability to hire the very best prospects from anywhere. As companies attempt to enforce return-to-office mandates, solid employees who otherwise would have happily stuck with their current employer will be available for more open-minded companies to snap up.

An Avoidable Handicap

The end result of a return-to-office mandate is crystal clear: companies serious about enforcing it are handicapping themselves. They'll likely lose some of their best employees while retaining their less-stellar hires. They'll find it more difficult and expensive to hire replacements since many applicants will reject an in-office job, and those who do accept it may expect moving expenses. Of course, the ever-growing cost of maintaining a traditional office space will be an ongoing drag on the bottom line.

Does this mean that every company with an in-office policy is doomed to fail? Of course not. Some specific enterprises require physical co-presence to perform at the highest level. Take marketing or advertising agencies or the "writers room" for a TV show, where the need for ongoing creative ferment may best be achieved by close staff relationships and those chance encounters in office hallways that could never happen on Zoom.

Accounting? Customer support? Even engineering design, which is largely done on computers these days? Not so much.

Is it hard to change your management style? Of course – but the job of executive leadership is to both build and lead the best-qualified staff to accomplish the goals of the company. As the necessary techniques to accomplish this change, it is absolutely critical for leadership to also change accordingly – otherwise they'll eventually get replaced, if not by higher authority (the board of directors) then by capitalist competition and Schumpeterian creative destruction.

Every company purports to want to be innovative; we all understand that the world is changing so quickly that it's impossible to stand still for more than a day, if that, and anyone that tries to stay put will not stay in business for long. Most companies, though, struggle with finding a definition of innovation that works for them, much less actually building innovation into their culture. Fortunately, we aren't asking you to hire research scientists to invent something completely new and never before seen: remote work has already been tried by others and found successful. Being open and willing to try new business models or strategies is itself innovative, if they're new to your organization, even though they've already been used elsewhere.

A remote-work model can produce material benefits for your organization—specifically, by allowing your employees to be more motivated and productive while saving money. It does not serve companies to claim to be innovative, and then reflexively demand for their workers to return to the office. As of this writing, Amazon is a notorious example: while being famously innovative and data driven, its blanket return-to-office order came without evidence of improved efficiency[5]— despite massive pushback from unhappy employees.

In reality, however, the challenge presented by the remote-work model isn't quite as monumental as it may appear. Leading a remote-oriented company certainly does have some differences to an old-school in-office one, but most of those are at the surface, not fundamental. At its core, a remote company is simply a new manifestation of what every modern company says, every successful company already knows, but far too many companies find extremely challenging: recognizing that its people are indeed its most important resource. And yes, I do realize that the previous chapter ended with nearly the same statement - that's to underscore just how vital a truth it is. In the next chapter, we'll start to dig in how to make that a reality.

[5] Towers, Jackie. "Amazon's Full-Time RTO Mandate: Perspectives and Data." HubStar. https://www.hubstar.com/blog/amazons-full-time-rto-mandate-perspectives-and-data/ Accessed Jun 30, 2025.

CHAPTER 3

PEOPLE FIRST AND PERSONNEL

Protecting and Growing Your Company's Assets

Over the course of the new millennium, companies have begun to use uplifting language to refer to their employees, such as calling them "partners," "associates," or pretty much anything other than "employees." HR also loves to stress that a company's employees are its most valuable asset.

Does anyone believe that's how the executives of Walmart, Amazon, and the rest truly feel, or even HR departments in general? Of course not: line employees are interchangeable cogs in a vast machine that barely knows they exist except for tracking their cost down to the penny.

If one individual employee "checks out" and does a bad job, or hardly works at all, how does the organization as a whole notice? That's right: it's their supervisor's job to watch and (hopefully verbally) "flog the laggards." In a traditional business setting, this develops naturally. With everyone in the office or on the shop floor together, human beings instinctively watch each other. Those with authority will object if they see something amiss, and those lower down will either obey or attempt to evade until they're winnowed out.

Managers who are accustomed to traditional office environments are understandably leery of work-from-home. How can they ensure that their people are working properly if they can't see them, or at least drop in on them unannounced?

Let's step back from this valid question and ask a different one: *does* every employee, in fact, have to be watched like a hawk to keep them from stealing, slacking, or sleeping? Of course not!

Human beings have a wide range of personalities. Some people will transgress even when they're being watched, and they generally end up in jail. Others will act virtuously no matter what, and we call them saints. Everybody falls somewhere in between. If you design your company around the assumption that you'll only hire saints, that's a failure waiting to happen. Even religious organizations, which explicitly have saintliness as an employment requirement, have a hard time living up to that expectation.

Even so, every office is staffed by people who are generally diligent and industrious – and not just the boss or founder. They may not be in high positions or receive much recognition, but everyone on the inside knows who they are and acknowledges (if only silently) that they keep the wheels turning.

Too many companies and executives may not act like they believe that "our people are our most important resource," but it's usually true anyway. Believing and acting on this core belief is always important, but it's absolutely essential for a successful fully-remote company.

—— Managers of One ——

If management can't keep a constant eye on everyone, how can they keep the business on course? The answer sounds trite: every employee, even the lowest ranking, must be a "manager of one." They themselves must be in charge of ensuring that they complete the work they're assigned.

Is everyone capable of doing this? No. Thus we see the first fundamental burden on executives of a fully-remote company: the need to be extremely selective, thorough, and involved throughout the entire hiring process.

This is contrary to modern HR hiring practices, where weeding through applicants is delegated to a specialized HR department that has minimal knowledge of the actual functions of the business or outsourced entirely to headhunters. How can you hope to secure the very best employees if most of the winnowing is done by people who do not themselves intimately understand what your business is all about?

To operate a successful remote business, hiring is not an occasional task left to department heads, nor can a siloed HR department handle it effectively.

Instead, selecting your next remote hire must be a high priority for senior leadership, right up to the CEO, because finding the right people is critical for success.

At OwnerRez, we rarely hired anyone without at least three interviews by leadership at various levels, including top-level executives. As of this writing, the CEO personally interviews all final applicants, no matter their job position. As a matter of policy, at least three fully-vetted candidates for each position are submitted to the CEO for consideration. Famously, Jeff Bezos's policy at Amazon required all final candidates to go through him, until the company's massive growth made that impossible.

Does this take a great deal of time and effort? Absolutely, and there's the ever-present temptation to "just get it done" and put a warm body in the seat. This is especially true for growing organizations that are understaffed and under-resourced, trying desperately to just plug holes as quickly as possible. Simply bringing in someone that can fog a mirror and plug holes sounds pretty good when you're having trouble keeping up with everything. That's a deadly mistake! The wrong person, in the wrong place, is worse than nobody in nearly any work situation, and at least ten times worse in a remote setting.

Interviews at fully-remote companies are almost never held in person. We use Zoom, Google Meet, or an equivalent because we need to be able to look the candidate in the eye. You'd also be surprised how much you can tell about them by their surroundings. Yes, often candidates use a virtual background rather than show their actual space, but even what they select can reveal something interesting.

In fact, a videoconference can actually be more effective than the traditional in-person interview, where all serious candidates know to present their very best, carefully-crafted personas. Remember the clichés about people who only ever wear a suit for funerals and job interviews? When you're trying to persuade someone to give you money on their own turf, common sense dictates that you be on your best behavior! But in your own chair in front of your own home computer, you tend to act more like your everyday self, for good or ill, than you act in the formal setting of an HR conference room.

So what are we looking for, beyond the obvious technical abilities required by the job? Character.

Diligence

Why do you do whatever your job pays you to do? For some people, it's just that: you do what's required to earn a paycheck. At quitting time, you clock out and go live your actual life, and that's fair to an extent. Not everyone is "always on," and we shouldn't expect that of most employees.

But what do you do when the boss is on vacation? Is that the opportunity to slack off and do the bare minimum to avoid being yelled at when they return? Or is that the opportunity to jump in wholeheartedly so the boss will return to a better situation than what they left behind?

For a remote employee, the boss is *never* "in the office." Choosing people whose approach is to lean into the work, rather than away from it, is essential.

This is a major reason for senior-level involvement at all levels of the hiring process, not just the very end of it. Especially for remote companies, having a deep knowledge of the organization's culture and ethos is critical to deciphering whether the potential candidate will "lean into the work" or "run away from it." Gauging whether an employee is the right "fit" from a temperament perspective is important at any company, but it's paramount in an environment where the boss is never in the office. Seasoned senior-level staff, by virtue of their years of life and leadership, can do so with a fair degree of accuracy, especially when several of them are involved in multiple interviews of the same candidate. The difference between a truly diligent employee and a timeserver is as different as night and day, justifying the investment of costly senior-level time.

You can gauge the potential diligence of candidates a variety of ways. First, a good application process should be quite involved – not in a tedious or exhaustive way, but in a fashion that corresponds to the position.

Wherever possible, standard aptitude test results can indicate relevant factors. With technical positions, this is pretty straightforward. Plenty of sites offer credible tests of expertise in .NET, CSS, and so on. If that's knowledge the job requires, no applicant should be offended. Soft-skilled jobs are harder to test for, but some assessments measure skills like attention to detail, problem solving, and reading comprehension. At OwnerRez, we used TestGorilla, but there are many other alternatives suited to different industries.

Another valuable tool is an extensive intake survey with freeform answer fields, customized for each job position, with prompts like, "Please tell us about your current work and why you are interested in another position." This is truly open-ended, and that's the point! An applicant who provides only short, one-line answers may be revealing that they're not particularly interested in the position.

Designing a quality questionnaire takes considerable time, and even more time to review applicant responses, but the payoff is well worth it. Are all the answers a simple "yes" or "no," or does a candidate provide more detailed answers? Is their writing clearly thought through and organized or a haphazard stream of consciousness? Did they understand the point of the questions and apply sound thinking or just bounce through and put down a simple answer?

One particular question that can generate revealing answers is, "Here's our website and support area: What do you think of it?" I've seen everything from the surface-level answer, "That's a pretty shade of green," to in-depth analyses of the content, the organization, missing content from the platform that could be useful – and yes, even the occasional typo that we missed. This is an effective gauge of the applicant's diligence and thoroughness and reflects how much effort they're willing to invest in working for you specifically, rather than "just anywhere."

Similar techniques apply during interviews. Rely on open-ended questions that force the applicant to think in real time so you can watch them process a genuine answer on the spot. Sure, an applicant might get help filling out survey questions or ask ChatGPT to do it for them, but they are considerably less helpful during a live interview. At the very least, live interviews still provide valuable insights into how applicants approach unfamiliar challenges, articulate their reasoning, and interact under pressure.

Since remote jobs rely on computers, this approach also showcases their general competence and familiarity with widely-used software tools. An applicant may not already be an expert with all the applications your company uses, but those with a wider array of computer experience will more quickly ramp up once they begin.

Mental Flexibility / Imagination

In the old economy, plenty of jobs seemed to require absolutely no mental energy. The job was literally just doing what you were told, tightening this bolt on the assembly line every 15 seconds, eight hours a day, five days a week, for 30 years.

As it turned out, this was a misunderstanding. Some manufacturing companies actually valued the input and intelligence of their line workers, even if those employees didn't have engineering degrees, and they ended up being far more efficient and successful than those who treated employees like robots. This was famously demonstrated in the latter half of the century by the companies that followed the management teachings of W. Edwards Deming, most notably Toyota and Ford. More recently, TV shows like *Undercover Boss* show how a CEO goes undercover as a line-worker and makes discoveries about their business that they would have never learned from remaining in the executive suite.

With remote work, the reality is even more stark. Working from home means unlimited potential for interruptions and distractions. That's why carefully selecting staff who display diligence is so important.

Sadly, so many people of some intelligence primarily apply their smarts to avoiding hard work. How much better if they instead used their brains to produce real benefit! Perhaps previously they worked in positions where their input or work ethic was not valued or rewarded. While everyone should take pride in their craft, it's difficult to be motivated to put in extra effort for the good of the company if you know you won't receive a raise, promotion, or even praise in return through the process.

As a leader, striving to identify and fairly compensate employees who engage their full brain power will result in a more productive team and the added bonus of discovering "diamonds in the rough." I'm thinking of several people who had previously been in fairly low-level, crank-turning positions for years before coming to work with us. We expected them to be solid work-ers, and they were. What we didn't expect was that, in an environment where new ideas are welcomed and leaning in is rewarded, they might excel even more than we originally expected. Both these employees and the company quickly benefited. What's not to like?

Not everyone needs to be Thomas Edison or Marie Curie to be successful, and that's fine. Everyone needs to have a willingness to constantly learn and to at least think through the inevitable problems that will arise. Some of this potential can be revealed by the intake survey, but it's more apparent during the interviewing process.

For example, for help desk roles, we asked candi-dates to answer questions that might have come from a client. Naturally, if they happen to know the right answer, that's a plus, but we don't really expect that. After all, they haven't yet been on the inside or under-gone staff training! No, the true purpose is to watch and evaluate their process of working through some-thing they don't know. Where will they look? Who might they ask? What sort of "Sorry, I just don't know" response will they come up with?

The truth is, no matter how knowledgeable or expe-rienced they are, prospects will *never* know everything. That's impossible, and we don't expect it. What we do require is for them to handle unknowns intelligently and professionally.

In an office, they might be able to walk down the hall and knock on someone's door for help. In remote work, we provide other tools like chat and email, but those might not always reach the right person promptly. Meanwhile, the issue is still alive and waiting. How do they handle it? Their answer reveals their intelligence, their diligence, and even their character, in one straightforward interaction.

The concept of "what do you do if you don't know" can be generalized to any job, even the most menial ones. How does the janitor handle a stubborn stain that doesn't come out with a common cleaner? It's often the difference between an adequate or bare-minimum employee, and a really good one. And its importance will only increase. How many jobs still exist where *nothing* out of the ordinary ever happens, and a veteran employee *never* asks for help or advice?

The Curse of Credentialism

What's the most common easy gatekeeping item HR uses to rule out an applicant? Their education level or specialized training. Maybe that made sense long ago, when a college degree or high school diploma meant something relevant. Those days are long gone. We've all read news articles about high school graduates who can't read their own diplomas and college graduates who have the attention spans of gnats.

Yes, certain occupations are legally mandated to require specific credentials. Is a computer science degree required, or even predictive, of a highly skilled software engineer? Wouldn't an experienced hardcore nerd who'd worked on his own projects since middle school likely have more raw expertise?

Using credentials as a filter for job applicants, beyond any legal requirement, is simply lazy. You're ruling *out* people with the actual skills you need, and you're likely including folks who majored in keggers and parties – the mere presence of a degree doesn't indicate either their diligence or their learning. That may be OK if you're hiring a social-media influencer, but not otherwise.

Look at the famous leaders and large organizations that have similarly realized that a college degree is simply not required for most jobs. Half of companies have removed this requirement[6] or are planning to, among them Walmart[7] and GM[8]. With a $350 million "New Skills at Work" initiative, JPMC[9] seeks to de-credential jobs that don't really require a degree.

You know best what skills fit the tasks you need done, not some college dean. Invest the time to understand what your best applicants are capable of, and expect impressive returns. Yes, this does put more

---

[6] Crist, Carolyn. "Nearly half of companies say they plan to eliminate bachelor's degree requirements in 2024." Higher Ed Drive. https://www.highereddive.com/news/nearly-half-of-companies-plan-to-eliminate-bachelors-degree-requirements/702277/ Accessed Apr 22, 2025.

[7] McGregor, Jena. "Walmart Plans To Remove College Degree Requirements From Hundreds Of Corporate Job Descriptions." *Forbes* magazine. https://www.forbes.com/sites/jenamcgregor/2023/09/28/walmart-plans-to-remove-college-degree-requirements-from-hundreds-of-its-corporate-job-descriptions/ Accessed Apr 22, 2025.

[8] Lutz, Hannah. "GM drops 4-year degree requirement for many jobs, will focus on skills." *Automotive News.* https://www.autonews.com/automakers-suppliers/gm-drops-4-year-degree-requirement-many-jobs-will-focus-skills/ Accessed Apr 22, 2025.

[9] Horowitz, Julia. "JPMorgan Chase is investing $350 million to get workers ready for the future." CNN. https://www.cnn.com/2019/03/18/business/jpmorgan-future-of-work-investment/index.html Accessed Apr 22, 2025.

burden on you as a hiring executive, but isn't that where such responsibility belongs?

—— Kids These Days ——

Business websites are full of articles appraising our younger working generations, commonly known as Gen Z. I don't believe I've ever seen a positive one. According to experts and experienced leaders, young entrants to the workforce are the dumbest, laziest, most oversensitive, least emotionally stable, least professional, least respectful, and just all-around the worst on record.

Strangely, when I was younger and reading then-prominent business periodicals like *Forbes, Fortune,* and *The Economist* (remember physical magazines that came in the actual mail?), they regularly published the exact same thing about the younger generation then, the ones who are now the thought leaders penning today's concerned critiques.

Are they correct? I don't know, and who cares? Unless you run the Army, it really doesn't matter if 50%, 60%, or even 90% of young people are worthless layabouts. Even if only 10% are any good – and I'm sure the true percentage is far greater than that – all you're really looking for is an infinitesimally tiny fraction of the many millions that percentage represents.

How many employees do you need? One? Ten? A few hundred? There are nearly 50 million Gen Z-ers in the U.S. workforce.[10] You *can* find top quality staff, and if you aren't, the problem is you and your process, not "lazy Gen Z."

"But all the young people I know are lousy!" Well, that's one great benefit of a fully remote company: you aren't restricted to just the potential employees that you know or that live within driving distance of your office.

Yes, there are certainly plenty of examples of young people who you'd be a fool to hire — so don't! That's why interviews are so essential to familiarize yourself with them as people. Regardless of age, if a prospect cares more about some pet cause than your company's core business, they're a better fit elsewhere. Let them go and wish them luck! Look for the first-rate candidates at the beginning of their careers, those with skimpy resumés who wouldn't make it past any AI filter but who bring the right character, intelligence, and drive to succeed. Their energy and goal-oriented efforts are likely exactly what your company needs.

As well as younger workers, OwnerRez also seeks experienced staff who have been burned out by the conventional corporate grind or found themselves unhappily pigeon-holed. Just because you've already been in the workforce for a decade or two doesn't mean all your depths have yet been revealed, sometimes even to yourself!

---

[10] Peck, Emily. "Zoomers will overtake boomers at work next year." Axios. https://www.axios.com/2023/11/22/gen-z-boomers-work-census-data Accessed Apr 22, 2025.

CHAPTER 4

TRUST BUT VERIFY
The Key Tenet of Remote Leadership

At this point, you may be scratching your chin skeptically: "This all sounds very abstract, even emotional, but I have actual work that needs to be done today and goals that need to be met. I can't count on magically hiring only people with 'the right stuff,' whatever that really means."

Of course you need reliable and productive workers! Even if you never see your employees in the office, that doesn't mean you can't hold them accountable. You just need to do so differently.

Consider reports of large companies firing remote employees[11] who used "mouse jigglers" and other such "cheat" devices to appear to be working when they weren't. This was framed as a knock on remote workers: "no in-office employee could ever get away with that!" But let's think about that for a minute. Anyone who's ever worked in an office knows some

[11] Kelly, Jack. "Wells Fargo Fires 'Mouse Jigglers' Taking Aim At Fake Work And Other Trends." *Forbes* magazine. https://www.forbes.com/sites/jackkelly/2024/06/18/wells-fargo-fires-mouse-jigglers-taking-aim-at-fake-work-and-other-trends/ Accessed Apr 22, 2025.

people work hard and others do as little as possible. It's human nature. Where you're located makes no difference.

What does make a difference is *effective* supervision. Consider the mouse-jigglers: someone using a mouse jiggler is not in fact doing any work, regardless of the "activity" their computer appears to show. Shouldn't that quickly become obvious? Aren't they supposed to be completing assigned tasks? What is their job, anyway?

If they are writing code, how many lines have they checked in? If they're making accounting entries, how many have they made and have they entered them on time? If they're compiling reports, those reports are being delivered on time or they are not. By definition, any kind of productive work results in an output that can be observed and likely measured, which is the job of a supervisor.

Viewed from this perspective, using a mouse jiggler is irrelevant. Was the employee's assigned work completed on time, accurately, and in compliance with expectations? If so, who cares about a mouse jiggler? And if not... still, who cares about a mouse jiggler? Not fulfilling the job's responsibilities is reason enough for sacking.

A company that has to fire people for faking work while doing nothing has far worse structural problems. For one, whoever was the boss of the mouse-jiggler-users should also be fired, as they are not performing their supervisory job. I'd suggest that *their* boss should also be sacked or at least thoroughly investigated. How can employees fail to do their work and nobody even notices? How can an executive not know that one of his middle managers isn't adequately supervising? Yes, mouse-jiggling is fraud, but the real problem is likely managerial incompetence.

Another possibility, particularly in a large company, is that employees outnumber the actual work to be done. In this case, some of the smarter people create "make-work" so they can *appear* to be busy, and in a sense they actually are. It's just not useful. Senior management should always be examining the work assigned on an ongoing and regular basis, not only to make sure it's being performed, but also to continually consider if it's still relevant to the company's current needs. Perhaps staff resources would be better allocated to other tasks.

Again, the true problem isn't that the line workers aren't working. It's that more senior employees aren't doing *their* jobs. The distinction of remote work versus office work is irrelevant.

For tasks that can be done remotely, ensuring the productivity of remote workers is fundamentally no different than if they worked in the office. In some ways, in-office fakery is *easier* than working remotely. We all know how to bustle briskly down the hallway, walking to and fro with pace to create an impression of frantic activity.

A boss who isn't capable of ensuring his remote workers are providing worthy value for their paychecks probably isn't that much better at managing employees located just down the hall. There's value in the maxim, "Expect what you inspect." Whether someone is in the next cubicle down or 10,000 miles away, good management requires regularly verifying what they're doing. If that's impossible, rethink the job description, if not the entire management structure.

In fact, if you are concerned about the potential for lazy and unproductive employees, managing remote workers is inherently a little *easier* because nearly everything they do will be documented electronically. In a physical office, while there's certainly plenty of computer work done using the same tools, there are also many other tasks that aren't recorded anywhere: making photocopies, setting up a Christmas tree in the company breakroom, and so on.

In a remote environment, equivalent work tasks are automatically logged. Instead of photocopies and mailed letters, emails are sent to a distribution list. That email's timestamp shows who sent it when and maybe even who *read* it when. Holiday greetings are likely sent using e-cards, which, again, travel by time-stamped emails. Accounting entries in QuickBooks Online? Edits to a Google Slides presentation or Google Docs marketing plan? Revisions to financial projection graphs in Google Sheets? All recorded. If that's not enough, computer monitoring tools are readily available that record what a computer is used for and when.

By now, most companies require employees to sign an Acceptable Use Policy statement, certifying that they won't use company computers for anything illegal or inappropriate and agreeing that the company has the right to monitor usage of equipment they paid for. As part of your company computer deployment, your tech staff can install a monitoring tool, such as Worktime.

Using this tool, supervisors and executives can tunnel down into their employees' workdays to see exactly what they were doing. When did they log on and off? What programs did they use? What web pages did they visit?

The Internet is infinitely large, but for most workers, the vast majority of their web usage will revolve around a limited number of sites: the company's own site, perhaps competitor websites, maybe an HR portal, perhaps sites related to their particular job, like a trouble-ticketing system. Unless their job involves marketing, visiting sites like Facebook or TikTok doesn't make sense.

—— Details, Details ——

So, how should you go about implementing a framework to track your employees' web activity when they're on the clock? In picking a computer monitoring tool, find one whose features suit your company's operation in a way that is both sufficiently fine-grained and appropriately general. For example, most of these tools allow the administrator to identify specific website domains as being work-related, definitely *not* work-related, or un-labeled and in need of managerial review. Will that same list apply to everyone in the company? Most workers shouldn't be spending time on travel websites planning their next vacation, but the CEO's admin certainly needs to be able to reserve flights and hotel bookings. Similarly, most folks shouldn't frequent Facebook or other social media during the workday, but the marketing team must.

A good tool will allow for customized profiles to apply to different workers or teams to make it easier to flag questionable usage.

Another problematic issue with website analysis is subdomains. Companies use these as a way to separate out different web-related functions. Consider Google: to use the search engine Google is famous for, you go straight to www.google.com. But Google relies on a variety of subdomains: Google Workspace's online documents are in docs.google.com. Email is at mail.google.com. Google's online calendar system is at – that's right – calendar.google.com. That's just the beginning. Some of these may be important to distinguish for your tracking system. If your company relies on Google Workspace, then the three subdomains listed above would be appropriate for all employees, but perhaps not news.google.com or photos.google.com. Maybe your work practices open all of Google's sites for workday availability. In that case, you don't want to manually add each subdomain individually to the whitelist. Instead, make sure your chosen tool supports "wildcards" like *.google.com (note the asterisk) that will cover them all at once.

Another concern is time zones and "official" hours. Some of the older work-monitoring tools were designed for jobs with a specific start and end time, like live call center operations, where it really matters if someone is 5 minutes late or leaves 5 minutes early. If that describes your business requirements, official hours are useful, but many remote jobs aren't anchored to specific hours, and flexibility is part of the job description. If that's your company, then you don't necessarily care when someone starts and ends or even if they make up missing hours on the weekend. All you care about is the work total. Not all monitoring tools effectively report that way, so be sure to check.

This problem becomes bigger if you have staff spread across the country or around the world. Some tools utilize a master company timezone, which means someone working on the other side of the world will end up with odd-looking reports. Whether or not you have geographically distributed employees today, you probably will eventually, so it's worth future-proofing your system.

—— Reality, Not Just Numbers ——

Most executives like to manage "by the numbers" because numbers suggest hard proof of what has been accomplished. If your financial numbers are accurate and show you're making a profit, the company must be doing something right, yes?

Not necessarily, as any number of financial disasters have demonstrated. Sometimes that's because of straight-up fraud, as with Bernie Madoff, Enron, and the FTX cryptocurrency scandal. Their numbers looked great until suddenly we all discovered that their financial reality was rather different and may as well have used Monopoly money in their accounting.

Other examples reveal that even when conscious fraud isn't the culprit, the numbers don't reveal the whole picture. For many years, Boeing and General Electric reported great earnings. More recently, they've struggled badly, leading many analysts to believe their past "profits" resulted from cutting essential investment in training, new technology, quality control, and experienced staff. Such actions can indeed goose your numbers in the short term but carry a far heavier price in the long run.

The same is true with measuring individual performance. Consider a help desk operation: what metrics are you looking for from your people? Is it the number of closed tickets? If so, you've simply created an incentive for tickets to be closed before they're really resolved. After all, the client can always just write in and open another one. Two hidden costs won't show up immediately in those numbers: annoyance to the client and wasted staff time as the second agent has to re-cover the first ticket's old ground. If the first ticket had remained open until the problem was truly solved, even if that "looked bad" on a dashboard somewhere, everyone would have been better served.

How about instead measuring ticket touches? That way, if a ticket requires several interactions to properly resolve an issue, the agent diligently working on it isn't penalized. There again, simply looking at that one number may camouflage an important reality. Suppose one agent takes three rounds to fully understand, diagnose, and resolve an issue, whereas a different agent can do it all in one round. We certainly want the first agent to persist and finish the job, but the second agent is clearly more knowledgeable and efficient. If your metric is ticket touches, the first agent looks misleadingly better because they're logging three times as many touches.

Am I arguing that numbers should be ignored? Of course not! Numbers, metrics, key performance indicators (KPIs): all these tools are absolutely essential to effective management, regardless of whether you're in an office or fully remote. They are not sufficient, however, and cannot represent the "final answer." KPIs are no more than warning lights on a control panel, pointing out where you need to investigate more deeply to truly understand what is going on and why. One could argue that this puts a heavier burden on a fully remote manager, but the burden is simply different. Offices establish methods to determine who is performing well and who is not. Without the convenience of physical proximity and personal observation, you may need to rethink how to effectively measure performance, but it can be done. More than that, though, taking the time to discern what you really want your team to do and the best way to measure their success provides an opportunity for general process improvement that's valuable however your company is set up. It's just more immediately essential in a remote organization.

In fact, if you are a traditional office-based business that's considering going remote, or at least making that an option, the effort spent in thinking through these issues won't only benefit the remote department. Once a business properly sets up the org chart, KPIs, job descriptions, and RACIs for remote work, those will work equally well for in-office workers, and the firm will benefit across the board. A well-monitored remote work operation won't just detect and prevent shirkers in their own homes, but in office settings as well.

—— Expect What You Inspect ——

Everyone wants their job to be easy, right? It's tempting to proceduralize the hard work of oversight: "If I look at these numbers and they meet these standards, all is well." As we've seen, though, that's only as true as your chosen numbers are relevant. There's another factor in play: your employees are not robots. They are thinking human beings who generally want to please their boss and be better paid. If they know you care about one number above all, so will they. That's why, for nearly all jobs, there simply *is no* "one number to rule them all."

Of course, some numbers are inherently important. If you run a widget factory, the number of widgets manufactured and shipped each day is always going to be top of mind, as is the number of returned defective widgets.

An effective leader will tunnel down into the numbers, perhaps in a slightly different way every time, to uncover new details and opportunities for improvement. Everybody in your organization should know what you deem the most important numbers, but also engage in regular analysis of other metrics and observations.

As we saw earlier, an effective remote worker must be able to discipline their use of time to efficiently accomplish whatever is assigned. Success means they have demonstrated detailed knowledge of their work and will be ready to discuss it on request. As a senior Ford executive said long ago, "Computers just make bad managers more effective."

—— Teamwork, Not Rat Finks ——

A principle of staffing I've observed is that generally "like prefers like." Good people prefer working with good people; bad people are embarrassed by the presence of good people and will try to make them bad or drive them out. Sometimes in business, you hear the saying "A's hire A's, but B's hire C's." It means that the best people want to work with other people who also are the best, but mediocre people feel threatened by the truly excellent and don't want them around.

This may seem like just another way of emphasizing the importance of hiring the best prospects to ensure a successful remote operation. That's true as far as it goes, but it also applies specifically to maintaining expectations, enforcing company policies, and preserving high quality.

Let's face it: we are all human beings who make mistakes. Every last one of us. We all need correction from time to time, or at least to have someone else point out an opportunity for improvement. That doesn't make us bad or lazy – it just means we're human.

Most people know where they rank on the totem pole. Someone who is lazy or ignorant probably knows that. The question is: do they intend to improve? If so, they'll welcome advice from more experienced people. Alternatively, those who would rather remain lazy and avoid notice may resent the pointers.

In any healthy work environment, particularly when working remotely, all employees need to feel empowered to point out anything they think is handled incorrectly. If they are accurately identifying a problem, that gives you the opportunity to take appropriate corrective action and prevent issues going forward. If they themselves are wrong, you have the chance to help them learn something they didn't know before.

As the leader, you may set the standard, but your eyes can't be everywhere; therefore, you should encourage cross-checking wherever possible. You may be thinking, "Nobody wants to feel like everybody is looking over their shoulder. I wouldn't want to work in an office full of narcs myself!" Of course not, which is why setting the culture and expectations is so critical.

A well-documented example comes from the airline industry, with its multitude of mandatory rules and regulations. If a pilot does something egregious enough, they can lose their license and their job. Fortunately, few offenses reach the magnitude of the "only an idiot would ever do that" sort. That still leaves a vast array of "that maybe wasn't the best idea" kind of errors in judgment.

Years ago, most airlines operated with the sort of hierarchy you'd likely expect: If someone senior is doing something, you just assume they know what they're doing and leave them to it. If they tell you to do something, even if it doesn't seem to make sense, you obey and maybe ask questions later.

Unfortunately, even the most senior person can suffer fatigue and make foolish mistakes. In the 1970s and 1980s, several spectacular crashes resulting in hundreds of deaths were caused by a senior pilot's actions that worried everyone else in the cockpit, but nobody dared challenge him.

In response, a philosophy generally known as Cockpit Resource Management was developed, where the head pilot is indeed expected to run the show, but the copilot and other crew aren't just minions to do what they're told. They are supposed to use their brains, offer feedback, and even push back if they believe something is wrong. Key to this system is the rule that the junior person asking a question won't be punished for the "disrespect" of challenging a senior officer. Obviously, this requires respectful ways to challenge a call, but the basic principle remains valid, and respect, politeness, and kindness are the foundation of any well-functioning office culture, whether it's online or in person.

It's taken several decades to largely embed this principle in airline culture. Even though some still resist the idea, we do see the benefits of less frequent major accidents. Every *honest* professional pilot will have a story about a time their brain just turned off and they were saved by a junior copilot speaking up. While the consequences aren't usually so dire, every honest executive will have a similar story, too.

Implement this sort of "let's fix the process, not the people" mindset in your operation, and your job of oversight can be vastly easier. It won't just be you keeping watch, but each and every member of your team.

Your immediate thought may be, "Won't that double the work if everything has to be checked by someone else? How is that more efficient?" Of course it isn't – and that's not what I'm suggesting. Think through the actual work being done by your people. Very little of that work is performed solely by one person without someone else reviewing it or seeing the results.

Software developers may write code alone, but regular code reviews by senior developers have long been a part of programming best practices. Most support issues will require at least a quick review of past interactions handled by other agents to properly understand the background of the question. These represent an opportunity to identify errors or omissions that could be improved internally. A fundamental principle of accounting is that no one person should ever be the only one with eyes on the numbers. Wise organizations create separate accounts-receivable and accounts-payable recorders, and they engage regular auditing by somebody else.

None of this has anything uniquely to do with remote work. Checks and balances are essential in any company, and so is continuous improvement. These long standing methods should be applied in any business, no matter how it's run. Unfortunately, they're often applied in a combative, recriminatory sort of way.

We've all caught a colleague making an error. What was your response? Did you try to quietly cover for your buddy's mistake? Did you use it as a shiv to stab a political rival in the back? Or did you work with them and with other subject matter experts to understand and address the root cause of the error in a collegial, helpful way, with the goals being to provide better service to your customers, personal improvement all around, and a more successful workplace? *That* is the sort of culture that every company needs, but a remote one most of all.

CHAPTER 5

LOVE AND LOYALTY

Practical Benefits and Win-Wins in Home Offices

I walked through the door of my house, greeted by the stagnant, hot, humid air of a malfunctioning air conditioner. Drat! Not only won't I sleep well, but my schedule might be upended for weeks. First, it may be a while before an A/C tech can be scheduled, and when they are, it'll be some unpredictable time on a workday. Someone has to be at home to let them in, of course, so that means a day off from work. They'll probably need to special order some part, which will take a while to arrive, and then they'll need to return to install it. Yep, another day off work. And that's the best case scenario. What about when the tech is delayed by an earlier job and has to cancel the appointment at the last minute?

We've all been in these situations. It's part of modern life. Every homeowner will encounter necessary repairs that can only be scheduled during workdays. Anyone with children will occasionally be awakened to a high temperature that requires them to stay home with a sick kid.

Traditional office environments have ways to handle this. You call in sick, taking paid leave time if you have it or simply losing that income if you don't. Everyone is used to this lose-lose situation: the company loses your valuable work without warning, and you either lose the income or the opportunity to use your time off for something more enjoyable.

One of the greatest benefits of a fully remote working environment is that much of the trouble and inconvenience of these situations vanishes, for both the company and the employee. With my malfunctioning A/C, I really only needed 5 minutes of not working to let the service technician in. He did his thing, and since I didn't need to watch him, I was free to work the whole time. As a fully remote worker, this frustrating event was invisible to the company, whereas in a regular office role, I would have been out and unavailable for potentially an entire day multiple times.

—— The Benefit of Flexibility ——

Other unscheduled work outages look the same. What do sick kids want to do? Sleep or possibly watch TV, which may require a parent to be physically present in the house, but this doesn't prevent working.

How about situations like doctor's appointments, where the employee has to travel somewhere and really can't work? They will lose a few hours, but most remote working environments involve solo tasks or work that isn't particularly time-sensitive. An employee can block off a few hours for the appointment, as they would for a meeting, and just shift their workday a little earlier or later. Both the work and the employee's outside obligations are managed in a timely fashion.

This offers a win-win situation since the employee's living conditions are vastly improved at no cost to the company. Need to pick your kids up from school? Sure, block off an hour in the afternoon and work an hour later to finish up what needs to be done. What's more, the inherent flexibility of most remote-work roles allows opportunities for those for whom in-person roles might be challenging. The disabled, the immunocompromised, those with serious family caregiving responsibilities that require their physical presence but allow them to still work: these prospects can provide your company a source of valuable contributors that's not available to your traditional office competitors.

—— A Job Worth Keeping ——

Let's face it, the classic stereotype of the greedy, grasping, skinflint boss who wants to wring every last drop of life out of his long suffering employees, paying them just enough to keep them from starving, didn't entirely perish with Ebenezer Scrooge. Most of us have had the misfortune of working for companies that are run this way, or at least have friends who do.

Set aside any moral or ethical issues with this approach. Success in the modern world of the knowledge economy depends on employees who have brains and know how to use them. Attempting to needlessly or unreasonably exploit employees can't help but backfire. If they don't feel appreciated and valued, the good ones will leave, and the ones that remain will likely not put forth their best efforts.

The improved employee lifestyle in a remote-work organization makes the most of this opportunity. The absence of a stressful and costly commute merely scratches the surface of the benefits.

Most employees do not live to work – they work to live. They have lives, interests, and obligations beyond what they do to keep bread on the table. When those obligations conflict with work responsibilities, stress results as they are forced to delay something important or juggle competing needs.

When a company can reduce this kind of employee stress, that's an improvement that doesn't show up on the balance sheet, at least not immediately. An employee who knows they can bug out for a couple hours to attend a PTA meeting, however, will be less likely to leave for a job that pays better but doesn't allow that type of flexibility.

Many of my hires left jobs that paid far more but kept them in the office for extreme hours, bracketed by backbreaking commutes. We all want more money, but if you never have time to enjoy it, what's the point? Somewhere along the line, these people took a hard look at their lives and decided that the tradeoff wasn't worth it. Instead, they brought their skills and experience to a company that was willing to work *within* their lives rather than to claim complete ownership of their schedules. These are some of the best employees because they know what the grass is like on the other side of the fence. They've been there, and they know that it's only greener because it's growing out of the septic tank.

Too few companies look at jobs this way. This makes those of us that do all the more valuable, far more than a mere salary grab for most people. Someone who's found themselves in such a position will be loath to abandon it, and they'll do their best to preserve it. Imagine what your company could accomplish if most of its employees were actually glad to have made sacrifices to work there!

CHAPTER 6

TOOLS OF
THE TRADE
An Efficient Software Stack that Works Anywhere

So you've decided to move your company, department, or team towards a fully-remote model. How exactly do you do that?

To be successful with any new system, you need the right tools, which in this context mostly means IT – software, hardware, networks, and the like. First, what work are your employees currently doing, and what tools are they using to do it?

Just about every modern office is already using many of the tools you'll need, like email and cloud storage. If your company operates its own data servers, you've likely already implemented secure remote access, perhaps a browser-based online login or a VPN. How do your executives access sensitive data from their laptops when they travel? What can they simply *not* do remotely, perhaps for security reasons?

Modern encryption and remote access protocols liberate us from a physical presence for computer networks, but the appropriate setup for each use case is beyond the scope of this book. Just recognize that it *can* be done and start exploring the appropriate research.

Most office workers already use conventional tools that, in principle, already work fine from anywhere. For example, email is likely already accessible on your team's personal smartphones. If you are using current versions of Microsoft Office, such as the 365 subscription-based suite, this can be accessed from anywhere using a browser or a virtual OneDrive installed on a laptop. This works with all Office365 products from any standard Internet connection. Another option is the Google Workspace suite, which not only includes Office equivalents like a word processor and spreadsheet, but also a video conferencing setup (Google Meet) and chat system (Google Chat). If you're paying Google to host your company email using their Gmail tools, you should have access to the whole suite at no additional charge.

—— Chat ——

An internal chat system is essential for efficient remote work. It functions as the logical equivalent of popping up over the cubicle wall or walking down the hall to knock on someone's office door. It's less formal and intrusive than a phone call and can be handled asynchronously. If the recipient has stepped away from the computer, the chat will normally stay flashing or highlighted until they can answer it.

While one-on-one chat sessions have their place, the true value of chats is seen in carefully configured group chats that allow discussions to take place both horizontally and vertically. Each team needs its own chat to encourage quick knowledge sharing and informal assistance. For example, if a support staffer is talking to a client with a question they can't answer, dropping a query in the chat may generate a quick answer that can be relayed to the client in real time. If nobody knows the answer, then that's a red flag for the team leader to research and perhaps request additional training from other sources.

Another useful chat group that's evolved over time has included the senior staff of all departments. How was this helpful? When a major event like an outage occurs, whoever encountered the problem first could immediately drop a message in the chat without chasing individuals down or sending an email that might get buried in an inbox backlog.

A somewhat more complex chat group arrangement involved the members of a given support level as well as the more experienced members of the next level up. That way, issues that might deserve escalation could be run by group members: Can we handle this at our level or do we need to escalate the issue formally to you guys?

In a sense, a well-used chat operation provides operational "grease" that prevents issues from getting bogged down in unnecessary red tape, just as walking down the office hall to knock on the door of someone in a different department can do. Of course, too much chat runs its own risks if threads become distracting or are hijacked. The obvious solution is to require that chats remain strictly business, but I don't think that's the right answer. After all, we *want* our remote employees to establish positive relationships with their peers, even though they may never meet them in the flesh!

One solution to offset this risk is an official "Water Cooler" chat that's used for everything *but* business. Yes, that's where you post photos of your pets, the pretty flowers you saw on your morning walk, or the scoreboard of the game you attended last night. Naturally, some people will use it more than others, but if everyone is a member of the chat group, they can all access the exchange and potentially learn something about their colleagues.

A key element to encouraging this type of interaction is the example of senior executives. If only the junior people ever appear in the "Water Cooler" chat, it may be deemed less valuable to your career, and possibly even harmful. If the CEO posts about his own kid's touchdown, though, that sets an example that will pay dividends in morale.

The size of your company may make a whole-company "Water Cooler" chat impractical. GM certainly doesn't have one and only one water cooler breakroom that literally everybody has to use. The principle is the same, though: personal interactions should be encouraged in remote work wherever possible, and chat is a highly effective, efficient, and affordable method.

A natural temptation is for the boss to join every chat. That's a mistake! No matter how cordial your relationship with your subordinates, do not expect them to say or share certain thoughts with you in writing. Healthy team-building and camaraderie require that staff can talk to each other privately or in selected groups without worrying about executive eavesdropping.

In a healthy culture, this is to your benefit. If someone says something that you really ought to know about, a colleague is likely to encourage them to share it with you directly, or they may bring it to your attention themselves (anonymized if need be, but at least you'll know about the issue).

Every organization will see its share of misunderstandings and hurt feelings. That's part of the human condition. Where it becomes a problem, though, is when they're allowed to fester and spread unnoticed. In a remote company, issues can go undetected for longer than in an office environment. We recognize when someone is disgruntled from their facial expression or body language when we pass them in the halls, less so on Zoom.

Intentionally creating opportunities for staff to talk to each other unmediated by higher authority helps bring issues into the light before they become serious problems. It may take a bit of effort and encouragement, but ad-hoc chat is a key tool for remote work. If you're using the Google Workspace system, Google

Chat is already included; many technology companies use Slack as a standalone solution, and Microsoft includes chat in its Teams offering as part of the Office suite. It's highly likely that you're already paying for one or more of these, so capture their value.

—— Office Document Tools ——

Historically, office productivity software was designed on the assumption that any given file existed in only one place at a time and was being worked on by only one person at a time. Even though corporate networks have existed for decades, one specific Word document or spreadsheet usually only had one person accessing it at any moment.

If more than one person tried, you'd end up overwriting each other's changes. Older versions of Microsoft Office handled this by creating hidden temporary "lockfiles" so that if you tried to open a file already in use, you'd see a warning message and be limited to read-only access. That simply doesn't work for the modern distributed environment. Microsoft knows this, and their most recent online version, Office 365, does a tolerable job of automatically updating multiple versions of the same file that are being edited by different people at the same time.

The program wasn't really designed for that in the first place, though, since the origins of Microsoft Office predate local office networks entirely. In contrast, the Google Workspace suite was designed in the era of the

Internet, so it's much more "at home" with this style of work. One particularly useful feature of Google Docs is that the document display shows everyone's cursor location, not just yours. You can see what someone else is looking at or editing in real time! This makes for smoother collaboration. For instance, you can create a meeting agenda document that all attendees can use, not just to add agenda items in advance, but to add notes, explanations, and action items that everyone can see and debate or clarify right then and there while the meeting is underway.

This is even more useful with spreadsheets, as selected cells are highlighted in separate colors based on who has clicked on them. If you're playing with financial projections, you can experiment with hypotheticals while minimizing the likelihood of stepping on a colleague's mathematical toes.

Unfortunately, the Google Workspace suite doesn't include all the features of the full Microsoft Office programs; however, they cover perhaps the most commonly used 90%, which for most workplaces is enough. Google apps generally can read and write documents to the Office specification. Again, a specialized or obscure Office feature may not translate perfectly, but the common operations will.

—— Video Conferencing ——

One stereotypical feature of a remote office is the Zoom call (or remote video-conferencing to use the generic name). Fortunately, the essential hardware and infrastructure has become all but standardized. Every modern laptop comes with a built-in webcam, microphone, and speakers. Desktop computers often do as well, and for those that don't, any missing pieces are easily purchased and connected.

For internal meetings, the built-in laptop cameras are generally adequate, but separate USB webcams are cheap enough that adding them to your standard equipment package for better picture quality may be worthwhile. One non-standard item that's well worth the price is separate lighting. Sure, you can select fancy "podcast" lighting systems, but even cheap clip-on USB lights can make a world of difference to video quality.

It wasn't that long ago that carefully staging the office wall behind you or investing in a dropdown green-screen was essential for presenting a professional appearance on video calls. Today, all the major video-conferencing software programs support smart-greenscreening, which automatically distinguishes you from your background. This feature places you in front of the virtual background of your choice without any special equipment. You may want to specify an official virtual background in your company colors sporting your logo or a virtual office with your logo on the (nonexistent) back wall. If it fits into your corporate public persona, I've found that allowing staff to choose their own backgrounds – within some common sense parameters, of course! – allows them to express a bit of personality. This lets them relate better to each other as well as to clients. I myself have a rotating folder of backgrounds – gardens, mountains, various offices, and the occasional cityscape – to add both beauty and variety.

As with the camera, laptop microphones are often sufficient, but a wide variety of higher-quality stand-alone USB microphones are available to improve sound quality. These sit in stands on your desk or are fastened to adjustable swing-arms. Our approach was to suggest a few options employees could choose from to upgrade their inadequate built-in microphones. Nearly all such microphones automatically configure themselves once connected, though some may need selecting in your video-conferencing software settings.

Speaking of that software: like Kleenex and Google, "Zoom call" has become a generic term for making a video conference call, but you have options beyond Zoom. They generally share features, but the most notable alternative is, once again, Google's offering of Meet. The big advantage of Google Meet in particular is that, unlike Zoom, you don't have to download any special software. It runs natively in the free Google Chrome browser that comes preinstalled on most computers. It also works on non-Google browsers, again without special software installation, though permissions have to be granted. This means that basically anybody in the world can join a Google Meet free of charge from any computer, even one in a library or Internet cafe where they cannot install software.

Google does have one limitation, however. To start a Google Meet call, you must use a Google account of some sort, whether that be your company's subscription to Google Workspace that provides your corporate email address or your own personal Gmail account. And, to accept a meeting invitation, you need to have an email address, though it doesn't have to be one provided by Google.

If you are using video conferencing to communicate with clients and business partners, not just internal employees, consider the tradeoff. Do you want to require employees to have a Google account, which is available for free from any browser? And, is it acceptable to require those outside of your company to install Zoom or similar software on their computer, which may not be possible if they're using a computer in an Internet cafe or library? This depends on who you're expecting to be meeting with, of course.

OwnerRez uses the Google option since we were already paying for Google Workspace, and we found very little client or partner resistance to needing an email address to join. I don't recall encountering anyone that didn't already have one, but then, OwnerRez is a B2B SaaS app; every customer we'd ever work with certainly has their own private computer and smartphone. Even job applicants all have at least a smartphone, and nearly all have their own laptop, which is typical for those applying for a fully-remote job. If they don't, they probably wouldn't be the strongest applicant.

Mutual Self-Care

We don't like to think about it, but because we spend a quarter of our lives working, we'll likely experience an urgent medical issue while we're at work. The same is true even if you're working from home, except that you're physically in the comfort of your own home. This presents a potential problem: if, God forbid, you have a heart attack in your office, someone should notice promptly and call for help. If you're alone at home, nobody may be around to do that for you. Worse, what if you're on a teleconference meeting, and your colleagues are stuck watching you suffer but unable to assist?

I can speak to this personally. One of my immediate relatives was alone working from home and suffered a severe stroke. As it happened, they were on a call with their boss, who quickly realized something was seriously wrong. As the boss, he had access to company records, so he was able to both send emergency services to my relative's house and call my relative's wife, who rushed home from her own office. The rescue squad arrived in time, and he's still alive.

Less happily, Zoom meetings have recorded assaults and even murders[12], where at least the video evidence aided investigations, but none of the horrified watchers could send help when it was needed.

---

[12] "Guilty verdict in horrific San Francisco murder recorded on Zoom." CBS News. https://www.cbsnews.com/sanfrancisco/news/guilty-verdict-in-horrific-san-francisco-murder-recorded-on-zoom/ Accessed Apr 22, 2025.

With a little advance planning, you can do better than to rely on the good fortune my relative experienced. One simple way is to use a voluntary emergency contact list on a secure shared document accessible by all employees. They can update their own information and instantly find contacts for their co-workers if needed. Naturally, some may choose not to share their full address, maybe just a phone number of their spouse, but it grants them the opportunity of doing something useful if an emergency arises. An active internal chat system, essential for efficient remote work anyway, also can allow quick access to a supervisor, who can pull more detailed information from company records should that be necessary.

—— Calendar Scheduling ——

One of the major advantages of remote work is its inherent support for flexible working hours. Of course, a major *disadvantage* is that you don't necessarily know who is working at any given moment. Some of the productivity tools we've discussed can help to an extent. For instance, tools like Google Chat and Slack can be configured to display whether someone is active, inactive, or logged out. For reasons unclear, though, I found these indicators not to be particularly reliable. A status could show someone was out, but if I sent them a chat, they replied instantly.

A unified calendar with shared visibility can also be helpful. Most online calendar tools allow you to configure your corporate calendars so anyone else in the organization can see them (though not necessarily edit). Each user can select from the complete organization list whose calendar events they choose to have included for viewing on their own calendar. For instance, I selected almost everyone in the company so I could see their calendars as needed. For most of them I un-checked them from my display list since I didn't need to actually view all their events all the time. The only actively-checked calendars were the dozen or so people I interacted with every day. Each of them were assigned their own color, so I could instantly see at a glance, "Oh, Fred is out today."

Maybe Fred was actually working today, but he was dealing with a complicated issue that required concentration without interruption. Integrated chat and calendar systems like Google's have a "focus time" event that indicates an employee shouldn't be disturbed even if they are present. Any chat messages sent to them will just pile up until the "focus time" event ends.

For remote-work online calendar coordination to work successfully, though, the personal responsibility for each employee to use and update their calendar has to be strongly embedded in the company culture. If you are going to step out for an hour to pick up the kids for school, that's fine, but make sure to block off that hour for all to see so someone isn't stuck waiting for you to respond!

By the same token, employees need to understand that a chat message is not an instant summons. We don't want people having to block off their calendar every time they step across to the kitchen to grab a cup of coffee. With clear calendar expectations, chat functions can be used flexibly. When you're out of the office, colleagues can still send you chat messages, recognizing that, no, you aren't going to answer them instantly, but you'll still see them and respond when you're able.

Naturally, people will use this flexibility differently. Some, like myself, choose an "always on" model and make ourselves available via chat nearly all of the time regardless of official hours. Many people find that unreasonable and impractical. If the calendar indicates that someone is flex-timing, perhaps working late because they were out all morning, they may respond to chats at hours they normally wouldn't; that doesn't necessarily mean they'll be available every evening going forward.

Whatever you decide is most suitable for your business needs and team dynamic, setting up appropriate calendar expectations is *vitally important*! The idea is to intentionally and publicly establish standards, as part of your company handbook or otherwise documented, that give staff the flexibility they need to live their lives, while ensuring both continuity and efficient work habits.

Time Zone Troubles

This brings up a related issue that can create big problems for a widely-distributed company: time zones. When your company is small and not too spread out, an official "company time zone" used for all meetings makes sense. New York or London time works well for that, since everyone in the world knows their time in relation to either the Big Apple or Big Ben. Generally, each employee can set their personal calendar to display in the timezone they actually live in, and the calendar system will translate automatically. If you're in New York and invite a Californian to a noon meeting, it'll correctly appear as 9 a.m. for them without any special action by either of you. You may also be able to set a secondary display. For example, my personal calendar shows New York time, but it's actually set to Chicago time because that's my home time zone. My calendar displays both.

Every major calendaring system includes a phone app, which makes keeping appointments so much easier! Most of us are already tied to our smartphones and their schedule reminders - which is why chaos can ensue when they don't work as expected. There's a very specific situation where this happens that a geographically diverse company needs to watch out for.

Typically, cellphones automatically update to a new time zone every time they connect to local tower as you move around. But this doesn't always happen if you're using WiFi rather than the normal cell network. This creates a potential gotcha when you're traveling overseas and don't have an international plan: a common solution is to just leave your phone in airplane mode to avoid connecting to the extraordinarily expensive international network, and turn on WiFi calling instead to use in the hotel and wherever else that's available. Everything will appear to work normally... except that your phone won't realize you're abroad in a different time zone.

You can manually set the timezone for your phone, and you'd think that would also automatically change everything that depends on time, like your calendar. Not so! At least, not for me the last time I tried this in Europe. My Google Calendar stayed stubbornly stuck in America, and I had to manually translate all appointments in my head – argh!

This worked well enough for a short trip, but find yourself at the wrong time of year and prepare for another quirk: Daylight Savings Time! Most countries don't observe this, nor do all U.S. states. Google Calendar knows who does and doesn't, and it can translate for you automatically when you're staying put, but only if it knows where you really are.

An even worse problem is date nomenclature. What's Christmas Day? December 25, of course, or 12/25 for short, right? Not in Europe, where it's 25/12, or 25-12. They place the day first and the month second. For Christmas, this isn't so bad because we don't have a month 25. Independence Day poses a problem though, because 4/7 could just as well be April 7 instead of the 4th of July.

It's even worse if you're using dates in spreadsheets because, while Excel can support all these options, importing a non-Excel file such as a CSV may operate from the wrong assumptions. The only solution I've encountered that addresses all cases is to demand that company dates always be expressed in full text: July 4 is impossible to confuse with anything else, even for a computer.

Establish standard practice as early as possible so it's baked into your company culture as a given. Undoubtedly, the major calendar apps will evolve and iron out the quirks eventually. Who knows, Congress may even decide to ditch Daylight Savings Time entirely! At least we aren't likely to use the Russian approach: no matter where you are in Russia, all official railway clocks resolutely display Moscow time, even 4,000 miles away in Vladivostok.

—— Cloud Storage ——

The Internet lets everyone work off the same data files at the same time rather than everyone using local copies that can get out of sync. We discussed this earlier under Office Tools, but depending on which tool you choose, you'll likely need cloud storage as well.

Google Workspace comes with a cloud storage system that can integrate with your own computer using Google Drive. Microsoft's OneDrive works similarly. Standalone cloud storage systems like Dropbox may work, but using the associated cloud storage that goes with your chosen office tool suite may be the most convenient.

While cloud storage tools are primarily oriented towards their provider's own file types, they aren't limited to just those files. You can upload anything to them. Most businesses use PDFs, for example, and everyone uses JPG, GIF, and PNG images. You might need a shared marketing folder with image files of your company logo and PDFs of marketing material for anyone in the company to access. You don't want everyone to *edit* those, though, so just as on a normal corporate network, someone has to control access permissions.

Data security is well beyond the scope of this book, but we can't avoid it. Set up properly, it smooths the path to effective collaboration, which is even more essential in a remote environment than in an office where, if worst comes to it, you can just walk down to Bob's cubicle with a thumb drive and copy the file you need. Set up incorrectly, it wastes everybody's time, opens the door to data loss and security breaches, or (most commonly) both.

As a side benefit, most cloud storage systems can record the history of who made what changes when. While this would be an enormous amount of data to actually review, revision history can prove very useful. If someone changed your file by mistake, you can review the history and revert to an earlier version. You can even use this information to prove that an employee was, or was not, working at a given time: if they said they spent all morning writing up the marketing plan, that's easily proven true or false.

If security permissions have been configured incorrectly, logs can show when someone has edited when they shouldn't have, allowing those changes to be reviewed and appropriate action to be taken. Sometimes detailed change logs are not active in cloud storage by default, so you or your account admin may have to activate them. That setting may even be hidden in the paid premium features area.

—— Agenda Tracking ——

How do you prepare and track agendas for your in-person meetings? There are as many different approaches as there are business consultants, from a standing whiteboard to a photocopied document distributed in advance. Like everything else in a fully-remote company, agendas have to be online.

One straightforward solution uses the online office productivity tools we've already discussed. For example, create an agenda document or spreadsheet and attach it to the meeting's calendar appointment. Using the document's permission settings, you can allow attendees to view or edit that document as desired. For recurring meetings, the same document can remain active indefinitely, with agenda items added and removed as needed, and perhaps a long-term reference section further down in the document for information you need to review frequently.

As meetings become more complicated – and particularly if you have formal minuting requirements – a specialized tool may be appropriate. We had success with an app called Fellow at OwnerRez. Like a bullet-pointed Word document, Fellow keeps track of agenda items. Unlike a simple document, Fellow lets you check off items as they are addressed, rolls unaddressed items up into the next meeting, and keeps a record of past agenda items that were checked off.

Since you're holding meetings online using a video conferencing program, you can record them. Fellow provides a text-to-speech transcription capability as well, which can be more efficient if you miss the meeting. Rather than listening to the entire recording, you can simply skim the transcript and jump to watch a specific spot of interest. Fellow integrates fairly well into the Google suite tools, though you can expect some glitches, particularly with recurring meetings. New versions of both should be available by the time you read this, so you'll want to test it yourself.

—— Webinars ——

Zoom calls and similar remote meeting tools are now a standard part of most white-collar businesses, not just remote companies. Similarly, "webinars" have become a well-accepted replacement for in-person seminars, training sessions, and other events that once were held in-person at vast expense. Because they have become so common, you'll find many resources available online and elsewhere discussing various approaches, methodologies, and tools. At its best, running a webinar can be like running a game show or talk show, where there's a live host interacting with a live audience. Unfortunately, most webinars more strongly resemble an HR briefing or 8mm movie from elementary school, where the biggest challenge is staying awake.

As with most public speaking, interaction is important, so choosing a webinar software product that supports interaction effectively is essential. At the very least, the tool needs to display a list of attendees, a running chat, a separate "questions" channel, and "raised hands" to grab attention. While you can both run the webinar software and deliver the content yourself, you'll see better results if someone other than the presenter operates the software. In fact, I found the most effective practice actually involves three people:

- The moderator/director runs the webinar software, controls what is displayed publicly to viewers, and perhaps introduces the presenter as an emcee would. This person needs tech abilities since they'll be the primary one to address technical problems like glitchy microphones and stalling video. They also need some public presence since they'll be acknowledging requests for the floor, introducing the next speaker, and so on.

- The presenter delivers the core presentation and interacts with guests or audience members who are invited "on stage" by the moderator/director.

- The chat moderator monitors the live chat activity between audience members, answering questions and providing references in the chat. While the presenter will likely answer questions too, think about how often a good answer includes a web link to a specific support document, article, or other website. Since the presenter is live on camera and can't easily chase down the right link, that's the job of the offscreen chat moderator. This person needs to be able to read and research fast, but they won't be talking or visible on camera.

You might wonder why the moderator/director and chat moderator should be two different people. If you have a small audience, perhaps one person can adequately perform both roles. I've found that if you have more than a dozen participants, though, and particularly if they are engaged and actively asking questions, it's difficult to listen to what is being said by the presenter, read questions as they come in, filter which questions go to the presenter, research and respond to other questions in the chat, and handle any technical or software issues, all at the same time live on camera. You'll deliver more professional results if you can split those two roles. Remember, the chat moderator doesn't have to appear on camera at all. They're just typing, so you should have an easier time finding someone who feels comfortable doing that.

Ideally, a webinar should include all the features we know and love from television and other media, from chyrons to fades, split-screens, and interstitials. Unfortunately, I haven't found any affordable webinar software that offers all this. High-end presentation software originally designed for professional videographers costs tens of thousands of dollars, and allows you to output to a streaming feed just like the big broadcasters. If you're open to that level of investment, that's great, but most businesses won't pay for that.

At the other end of the scale is the more afford-able software designed for webinars specifically, as opposed to broadcast TV. Unfortunately, none of these offer the full breadth of features that you may want, so whatever you choose will involve tradeoffs. Crowd-cast is a well-known entrant in this field, and it works tolerably well, but with annoying limitations. Oddly, at least as of this writing, the app doesn't even offer such common functions as a built-in automatic virtual greenscreen for participants to select their own back-ground! Even the free video conferencing tools can do that. It does however handle interactions with attend-ees well.

Due to the inherent interactions with some of the more complex bits of your computer – live video from the camera, live audio from the mic, stream-ing data in both directions – webinar software can be both complex and finicky. Plan on setting aside time to test out several options to determine which set of features comes closest to meeting your needs. Then experiment to ensure you can use it reliably, such as at an internal training session where all attendees are employees and any problems won't become a public embarrassment.

Even after your internal test, each and every webi-nar needs a "green room" session immediately before the show starts, just to confirm that all the gadgetry is still working properly. The same presenter who's suc-cessfully facilitated dozens of webinars on the same computer hardware may find they need a reboot to reactivate their mic. Make sure to allow at least 15 minutes in advance for the major participants to join and iron out any glitches.

The above warnings and caveats may sound discouraging. Don't despair! Webinars are actually extremely powerful and time-efficient, well worth your investment in research and learning upfront. What's more, they do not require vast sums or a full-time production team to succeed. I've attended webinars produced by billion-dollar companies involving C-level executives that aren't as good as videos their teenage kid posts on TikTok.

You can do better! Maybe you won't meet broadcast-news levels of technical excellence, but nothing's preventing your webinars from looking professional, and you don't have to break the bank.

—— Telephones ——

As of this writing, we're coming up on the 150th anniversary of the invention of the telephone. To think we complain about limitations of desktop computing that date back to its origins a mere four decades ago! Yes, the global telecommunications network has seen improvements that Alexander Graham Bell could never have imagined, but the most basic elements aren't all that fundamentally different from what he and Thomas Watson used for their very first call. It should be no surprise, then, that phone calls don't always play well with the modern Internet.

We've all encountered VOIP telephone systems. That stands for Voice Over Internet Protocol, which does exactly what it sounds like, turning your computer into a phone using the Internet. VOIP software presents a telephone dialer on your computer screen, allowing you to place an ordinary call to any telephone in the world using your computer microphone and speakers. This capability can be configured to allow your employees to place calls from wherever they are, using your company's telephone number as their caller ID. This is useful since many people won't answer calls from numbers they don't recognize. It's also generally a bad idea for employees to make work calls using their own personal phone numbers because customers can hound them personally thereafter. Even if you issue company phones to all employees, that still presents problems. You generally don't want customers to call one specific person's number rather than a general number that can be routed to whoever's on duty.

A more sophisticated central VOIP system can do this, tracking who is logged into their computer, not engaged on another call, not in a scheduled meeting, in the desired department or workgroup, and so on, and ring their virtual phone in their browser window or app. This can be scaled to any level. Major airline JetBlue, for example, has a remote call center team of thousands of employees that handle traveler issues by phone, email, and social media, all synchronized together and mapped to a single customer account.

Another option is what many tech startups are doing: not have a live incoming phone number at all. OwnerRez had a phone number, but calls normally went straight to voicemail, which automatically created a trouble ticket for the helpdesk staff. This could be done by email if our database recognized the calling phone number, or by an outgoing VOIP call placed by the agent that drew the ticket.

CHAPTER 7

YOUR SECRET WEAPON

How to Beat Your
Competition's P&L

What does your business spend the most money on, i.e., the largest single budget category in your profit-and-loss (P&L) statement? To some extent, that depends on the nature of your business. For a bulk wholesaler, "cost of goods sold" may very well be number one. For service businesses, though, the largest costs typically revolve around the business' employees. You're familiar with the obvious subcategories: salaries, health insurance, and ancillary costs like various employment taxes. Some other costs aren't labeled as employee related, but inherently are, like their office space.

Also consider additional costs or doubled-up costs that make the true cost of an employee even higher than it seems at first glance. For example, an HR staffer is a people expense since they're an employee earning a salary, but the reason that job exists is because the *other* employees exist. If you add additional employees for any reason, sooner or later you'll likely add another HR position as well.

What if you could significantly cut back on these costs while improving quality? No, I'm not talking about replacing half your employees with AI chatbots and the other half with robots. A fully-remote company provides the opportunity for just such a cost savings, while at the same time improving quality, productivity, and job satisfaction.

_____ The Rent is Too _____
Damn High!

First, let's address the obvious: if your employees are all working from home, you eliminate owning or leasing office space for them. All by itself, that can make the difference between profit and loss. The actual cost of office space naturally depends on location, but the lower end has been estimated at around $5,000 per employee per year, going up to $20,000 or more in larger cities. That's just the cost to the company of providing the space itself and doesn't include any allowance for time and expense in commuting, as that's borne by the employees. Office furniture and equipment adds to this cost, though computers are required by both in-office and remote employees.

The question for you as a business leader is: what could you use that money for that's more valuable than a physical office? Or put differently: does a physical office provide that much value versus whatever else you might spend the money on?

Everything in business (and life) involves tradeoffs: money and time invested in one thing is money and time *not* invested in something else. We can fall into the trap of assuming that "it's always been done this way" means that it always *should* be done this way. Perhaps the reason for a particular business practice is now out of date.

A century ago, every office was well supplied with low-level clerks and secretaries, whose job was to generate and handle paperwork. Executives needing to write a report or letter would dictate it to a secretary, who'd then pound away on a manual typewriter to produce the letter and drop it in the mail. Similarly, offices were equipped with mailrooms, where incoming mail was routed manually to the appropriate internal office via a young fellow pushing a cart around, dropping off and picking up envelopes as he went.

All of that is nearly gone now, not because businesspeople no longer communicate, but because we've developed better ways that don't require specialized workers. Modern computer keyboards can be used by anyone, and for decades now, students have been taught how to type. For those that would rather not, they can use text-to-speech apps on their smartphones. We can read emails from anywhere and instantly forward them on to someone else. We don't need teams of staff schlepping physical paperwork around.

Of course, paperwork hasn't vanished altogether. Legal requirements and archiving rules necessitate hard copy files, so truly paperless offices are few and far between. It's a small fraction of what it used to be though. What happened to all the money once spent on secretaries and mailroom staff? It went somewhere else, and so did they. The evolution of society took a few decades to work these changes through every workspace, but over the years, retiring secretaries weren't replaced.

We are experiencing a similar effect in office space, although we will undoubtedly always need physical offices. Sometimes this will be required by law. The Department of Motor Vehicles isn't likely to go virtual anytime soon, nor are courtrooms. Other times, the nature of the work really does benefit enough from having everybody physically together to justify the expense. Pure creatives like fashion designers might be an example. Any company leader needs to ask themselves: what corporate functions *truly* require, or at least benefit sufficiently from, physical co-location? Each individual work role needs to be examined in that light.

—— The Best Hire Possible ——

So you've decided that you need to add someone to your team in the office. You've prepared the job description, posted it on your website, turned the #Hiring hashtag on your LinkedIn profile, maybe posted it to Indeed or one of the other help-wanted sites. You might even have posted a physical "Help Wanted" sign in your window, if you have one.

Who's going to apply for your position? Some will see the job opening or have it pointed out to them by someone else. Beyond that, if your job is onsite, only people who live fairly close to your office or are willing to relocate will choose to apply.

If your office is in New York, don't expect many applicants from California or Texas. You might see a candidate from Chicago or Boston, but that raises another question: are you willing to pay relocation expenses? Sometimes, for a really top-tier candidate, this extra cost is worth it, but you accept increasing the risk with every new hire. You never know *for sure* if someone will fit the role, and if they don't, you're out the relocation money.

What are the odds that the truly best hire is within commuting distance of your office? Mathematically, they have to be low because the overwhelming majority of potential candidates live somewhere else. For a fully-remote company, these candidate pool limitations vanish. You are fishing in the largest possible pond, consisting of everyone in the world – or at least, everyone on the Internet who speaks your language and is within reasonable time zone range. This allows you to be far more selective than you could ever be for an onsite role, even in a major city with millions of potential nearby candidates. More than that, though, you can open your doors to candidates who could not take a job in your office unless they already lived in your city within commuting range.

—— A Ramp Upwards ——

Back in the early 1990s, a signature achievement of the George H.W. Bush administration was the Americans with Disabilities Act. Among other things, the law banned broad-based hiring discrimination against the handicapped. As with most legislative action, this turned out not to work as advertised, and for a perfectly obvious reason: many disabilities legitimately impact someone's ability to do a job. For example, someone confined to a wheelchair won't make an effective UPS delivery person. In a nod to logic and common sense, the ADA only requires companies to make *reasonable* accommodations to allow a disabled person to do a job, not unreasonable or impossible ones.

Despite these well-meaning attempts, for decades after the ADA was passed, the employment percentage of disabled people decreased[13] - the exact opposite[14] of the intention. Only in the early 2020s[15] did employment of the disabled start to increase. What might have hap-

[13] Miller, Tracy. "Why Are People with Disabilities Underrepresented in the Workforce?" Discourse. https://www.discoursemagazine.com/p/why-are-people-with-disabilities Accessed Apr 22, 2025.
[14] Burkhauser, Richard and Daly, Mary. "The Declining Work and Welfare of People With Disabilities: What Went Wrong and a Strategy for Change." AEI Press. https://www.researchgate.net/publication/265245074_The_Declining_Work_and_Welfare_of_People_With_Disabilities_What_Went_Wrong_and_a_Strategy_for_Change Accessed Apr 22, 2025.
[15] Andara, Kennedy, et. al. "Disabled Workers Saw Record Employment Gains in 2023, But Gaps Remain." Center for American Progress. https://www.americanprogress.org/article/disabled-workers-saw-record-employment-gains-in-2023-but-gaps-remain/ Accessed Apr 22, 2025.

pened around that time to cause this? One answer is clear: the COVID pandemic drove most onsite workers out of their offices and had them working from home instead. Business carried on, including resignations, firings, and hirings, but now all this was conducted remotely, too. In these circumstances, many disabilities didn't matter. A wheelchair-bound person will naturally have modified their home to accommodate their needs. If they're working from home, the company bears no additional cost or inconvenience. Since everyone in a Zoom call is normally seated anyway, the company might not even know of a disabled applicant's condition unless the candidate chooses to disclose it.

Yes, it is possible for companies to design specialty workspaces for people with special needs at vast expense. *Fortune* magazine recently discussed one company that configured half their floor to suit the "neurodivergent"[16]:

> One side of the office has traditional white overhead lighting, while the other side has more muted yellow lighting. Throughout the day, the fixtures automatically adjust their brightness in response to the daylight. The office also has different zones with designated thermal controls, so employees can work in the temperatures they prefer.

[16] Burleigh, Emma. "The rise of the neurodivergent-friendly office—How a once-niche workplace idea is catching on in corporate America." *Fortune* magazine. https://fortune.com/2024/12/11/neurodiverse-worker-office-accomodation-gaining-popularity-workplace-architecture-interior-design/ Accessed Apr 22, 2025.

Understood's office also has quiet rooms for employees who are distracted by coworker chit-chat, in which staffers are not allowed to speak or play noise. For people with hearing impairments, the floor has an assistive listening system that connects to cochlear implants; this technology transmits audio directly to their compatible receivers, allowing them to more easily listen to conversations, meetings, or presentations. Employees with visual impairment have access to the name of each room written in braille on the wall outside.

The office has a variety of tables and chairs to accommodate different body types, abilities, and working styles, including cushy white stools, linen-bound rolling chairs, plush elongated couches, and wooden chairs. The desks are also adjustable – employees can choose to work standing up or sitting down. When it comes to soundproofing, some rooms have squiggly and colorful chunks of foam hugging the inside walls to give them extra insulation. Some specific furniture, like grey loveseat pairs shaped like cocoons, provides a sound shelter for staffers who want to talk more privately.

From the description, one can only imagine the complexity involved. How many different types of chairs must be purchased and maintained? Modern LED bulbs support a fair breadth of brightness and color-tone configuration, but does that suggest that everyone has their own private office? Or do people sort themselves into bullpens on the basis of color temperature? What happens if office-mates agree on the light color tone but disagree on the thermostat setting?

Remote work eliminates all of this managerial and financial overhead while permitting an unlimited array of possible accommodations. Each employee can design their own workspace in their own home to suit their preferences, be they medical, familial, logistical, or nothing more than personal taste. Their choices won't affect any other co-worker in any way or cost the company a dime.

In short, remote work is a tremendous boon to people who, for various reasons, could not or would rather not travel to an office, but who are perfectly capable of performing productively from the privacy and convenience of their own homes. The physically handicapped are the most obvious example of this, but far from the only ones.

Another largely untapped source of work-from-home employees are those who have family caregiving responsibilities that require their physical presence at home, but not active caregiving at all hours of the day. Consider someone whose elderly parent is suffering from mild dementia. The elder can't be left alone, and memory-care facilities are extortionately expensive. Often just having a loved one present in the house, even in a different room, is all that's required.

Someone in this family situation can't be away from home for 10 hours a day of commute and in-office work. Nothing prevents them from working a full day from home, though. We're not talking about insignificant numbers here either. According to[17] the Duke University School of Nursing and the AARP, "17% of the country's adult population cares for an adult over the age of 50." Imagine the benefits of including these potentially perfect prospects in your search pool!

_____ Opportunity Knocks _____ on Obscure Doors

A common topic in public policy circles is the concept of the "digital desert." This refers to places off the beaten path, a long way from major cities, where good-paying (or any) jobs are hard to find, and Internet service is spotty at best. In the three decades since the dot-com era, vast sums have been spent to close this gap. Today more than 90% of Americans have access to broadband Internet service[18] where they live, and that increases every day.

Many young people who grow up in rural areas choose to move to bigger cities where they perceive better opportunities for career advancement. Not all do, though, and many who do make this move would prefer to stay closer to home, friends, and family if they could. For every would-be New Yorker, there's someone who feels more at home on the range.

[17] "Adult Children Pick Up the Responsibility of "Aging in Place" Parents." Duke University School of Nursing. https://nursing.duke.edu/news/adult-children-pick-responsibility-%E2%80%9Caging-place%E2%80%9D-parents Accessed Apr 22, 2025.
[18] "FCC: 24 million Americans still lack high speed internet." CNN. https://www.abccolumbia.com/2024/03/15/fcc-24-million-americans-still-lack-high-speed-internet/ Accessed Apr 22, 2025.

This presents two great opportunities for a fully-remote company. First, those people who have chosen to live in rural areas but still have reliable Internet access are a largely untapped source of quality workers unavailable to your in-office competition. Even more, being fully remote allows workers to choose where they want to live, which may not necessarily be where they live right now. You might see them as a New Yorker, and legally that's what they are, but that may very well be simply due to expediency.

I recall one applicant who applied, interviewed, and accepted a position from Brooklyn, but by the time her first day of work arrived, she was living in Florida. Making this choice added greatly to her satisfaction, quality of life, and happiness with her job – at absolutely no cost whatsoever to the company. She relocated herself without telling us, at her own expense and for her own benefit, but our company's remote work policy made it possible.

——— Work Never Sleeps ———

Decades ago, most businesses shut down at night. Even large companies couldn't be reached after hours: call back during the day, please, or write a letter. Those days are long over. Even today's small companies are expected to make someone available – if not on the phone, then at least online, and if not fully 24/7, then at least well into the evening. In part, that's because people work long hours and many have side hustles, so the only time they can reach out to you is after normal business hours.

Email communication helps with this. Just like mailing a letter in the old days, it doesn't matter when it's written or sent, it'll be waiting for you in your inbox whenever you're available. Even so, customers now aren't willing to wait indefinitely for a response. They prefer an answer *right now*, or as close to now as humanly possible. In the Internet age, instant gratification simply isn't fast enough. Knowing this pressure, dutiful staff may find themselves working longer and longer days to help "just one more" customer so they don't have to wait until tomorrow, or, God forbid, until after the weekend. Bosses may encourage or tolerate this, since on its face it's good for customer relations. Give the customer what they want!

Since customers want help at all hours of the day, night, weekends, and holidays, the problem becomes *never* having a time when work is actually finished. You'll always have "just one more" thing you could do to help somebody instead of going home to have dinner and to sleep. For employees for whom a job is just a job, that's not a problem: when 5 o'clock comes round, it's quitting time! Off they go.

As we've already discussed, a remote-work environment makes it both important and possible to recruit diligent employees who truly care. This presents a problem, though. You *do* want them to care, but you *don't* want them to burn out through overwork, not just because you (hopefully) care about them as human beings, but also because that's bad for the business. We've all encountered burnt-out workers who simply don't give a hoot anymore. People like that certainly aren't going to make your customers want to *stay* your customers.

Extending your business hours brings challenges of its own. Most people prefer to awaken in the morning and go to bed at night. Not many people want to live the other way around. That's why the night shift generally pays workers extra, and the graveyard shift even more: more money is the only way to entice people to work then. Even so, working odd hours can cause all kinds of problems, from heart issues[19], to the obvious logistical problems with personal relationships, to even such mundane concerns as how to attend doctors' appointments if their office hours fall when you're normally sleeping.

A remote-work environment offers several ways to address this effectively for little to no cost. Since you are no longer restricted to hiring employees within driving distance of your office, you automatically have a wider range of time zones. If most of your staff happens to be on the U.S. east coast, simply hiring someone in California will automatically provide the option of a "shoulder" shift. Their own "normal" work hours are three hours out of phase anyway, and you could probably nudge them an hour or two further without much difficulty.

In the other direction, Ireland is 5 hours ahead of New York time. That's why so many tech companies

[19] "Working the night shift comes with many struggles." UCLA Health. https://www.uclahealth.org/news/article/working-night-shift-comes-with-many-struggles Accessed Apr 22, 2025.

base a support team there. The language is English, the culture is not too different, and the time is just enough out of phase to be useful while still allowing some degree of team overlap. Add Hawaii to the mix, and now you can cover everything except the very dead of night without requiring any individual employee to work truly weird hours. Hawaii, California, New York, Ireland – but what about tapping the rest of the world for remote employees so work can continue 'round the clock? That's a larger issue worth its own chapter.

CHAPTER 8

OUR FRIENDS ABROAD

Future-Proofing in a Changing World

The last decade of the twentieth century and all of the twenty-first so far has seen the world becoming both much smaller and much larger. It's smaller in that it's become ever-easier to conduct business with anyone in the world. Once limited to fax transmissions and expensive overseas telephone calls, we can now all communicate instantly, for free, with full live voice and video from any computer or smartphone with an ordinary Internet connection. By the same token, the world has also become larger in that we're no longer just competing with our counterparts in the same city, state, or even country, but anyone in the world, both as employees and as companies.

No doubt many reading this book share the concern: "If my company went fully remote, why would they want to keep me at American or European wages? Why wouldn't they just hire someone in India, China, or the Philippines for a quarter of the salary?"

As we've seen over the past few decades, many companies have already gone that route, and it seems to work for a while, at least on paper.

Other factors add expense. If you have a physical office in another country, you have to deal with the legalities of creating a foreign subsidiary, obeying foreign laws and regulations that might be wildly different and unfamiliar, and any number of other potentially costly "gotchas." With work-from-home remote employees, though, nearly every country allows "consultants" who are paid with the local equivalent of the American Form 1099. Their "employer" doesn't have any legal local presence at all – no business license, no taxes, no filings of any kind. They just send money, and the worker who receives it is fully responsible for paying whatever is required, just as American 1099 consultants are.

Examining the history of offshoring shows that, while it can be a valuable technique, it's not the panacea many assumed it to be when the trend began. First, overseas wages have been increasing exponentially. In many countries, they're still noticeably lower than U.S. wages, but not nearly as much so as twenty years ago. The wage gap has been closing slowly but steadily, and it's easy to imagine a not-too-distant future when a worker of equivalent skill, experience, knowledge, and productivity in Bangalore will command close to the same paycheck as someone living anywhere in the U.S. other than a primo metro area like Silicon Valley.

For the very best tech workers, like top-quality senior software developers, that's already largely true, particularly when adjusted for local purchasing power parity. In positions requiring less expertise, the wage differential is still significant, but it's closing nonetheless. Some of this is caused by paying American workers less, with all the political controversy this creates, but most is simply foreign workers demanding more compensation over time as their nations develop, just as we've seen everywhere throughout history since the dawn of the Industrial Revolution. While offshoring lower-level roles still can save money, those savings have shrunk over time and can be expected to continue to do so.

Penny-Wise Pound Foolishness?

Not too long ago, I had lunch with a senior executive of a large publicly-traded company to discuss how our respective companies handle customer support. OwnerRez has always relied primarily on U.S.-based employees, adding overseas workers only for specific reasons, and never with the intention of saving money by paying lower wages. In contrast, my friend's company actively sought to move work overseas as much as possible just to save money. As he put it, "The stockholders expect us to be hiring people wherever the prevailing pay is the least. If I don't do it, they'll find someone else who will."

This perspective isn't uncommon, particularly in publicly-traded companies whose executives live by the stock market's reaction to their quarterly earnings report. At the same time, those who experience a giant company's tech support through an overseas call center – well, I hardly need to describe the well-worn stereotypes about customer frustration and dissatisfaction.

How successful were my friend's efforts? I'm not privy to their internal company statistics, and being so much larger than OwnerRez, their operations aren't completely comparable. My back-of-the-envelope calculations, however, suggested that by adjusting for the higher number and size of their clients, our American-based support was between five and ten times more efficient considering the number of staff versus client support provided.

Does this mean that our support must therefore have been worse, since we had proportionately so many fewer staff? No, Gartner rated OwnerRez as having the best support in the industry. I have no doubt that our support staff were paid considerably more than the overseas staff of my friend's company, perhaps even three or four times as much, but if they are between five and ten times more productive, isn't that a cost savings? The overall quality of results would appear to be better, too, based on our higher support rating by outside research agencies.

The point I'm making is not that all overseas employees are bad, incompetent, ignorant, low-quality, or any other negative characterization. That's flatly false! I've worked with individuals from every corner of the world who are at the very top of their areas of expertise, the best of the best. So has anyone else who's been in the tech world for any length of time.

I think the true problem is the initial attitude going into the hiring and buildout process. If the overarching goal is to save money in the near term, your natural inclination can't help but lean towards paying less, which almost guarantees you will *not* get the most capable, most experienced, and most qualified people working for you *anywhere.* In most technical fields, the productivity, quality, and efficiency of the very best people is so much greater than that of the average or low-end that even though they are making far more money, they are still the most cost-effective when you analyze their results.

—— A Solid Foundation ——

What does this have to do with how a fully-remote company environment promotes offshoring? Historically, most companies that have pursued offshoring already have physical offices and infrastructure. Their whole culture is oriented that way. This makes sense because until recently, that's been the business status quo. Company operations, at their core, follow the classic model: expect employees in the office, schedule in-person meetings, and so on.

Because of these traditions, most offshored operations involve a fair amount of siloing: people work primarily with other people in their own office, and interactions with people elsewhere are limited. Modern technology eliminates such limited interactions since we all have phones, email addresses, and video-conferencing capabilities. It's more a matter of established culture and habits.

Since traditional offices largely operate with their own local office culture, when clients regularly interact with multiple offices from around the world, they may not receive consistent client support. This is widely known in the world of tech support. With many global companies, you're better off calling at a particular time of day to be routed to a particular global office that provides better support than offices elsewhere.

In contrast, when a company is fully remote, this siloing and its resulting inconsistencies are less of an issue. Individual workers can communicate with whoever they need to. Over time, they'll establish closer relationships with some co-workers based on any number of factors, but *not* physical location. Yes, differing time zones and non-overlapping work hours limit synchronous interactions to some extent. At the same time though, a culture of allowing maximum work-hours flexibility aligned with productivity will expand the potential overlap. This supports improved knowledge exchange via unofficial and informal channels, which enhances the effectiveness of formal knowledge exchange programs.

A company that's oriented around the expectation that everyone is remote will be better prepared to accept employees all around the world, whenever that becomes appropriate. Since managers will already be adept at managing remote employees, the geographical distance apart will be irrelevant. You'll also have the ability to stretch your global company muscles slowly by degrees, one new hire at a time, rather than opening a completely new foreign office all at once, with all the startup and learning costs that entails.

One other important element is worth a mention. We all know the variety of accents and dialects that can affect understanding. A fully-remote operation will have its share of live conversations and meetings, but I've found a greater reliance on text communication: emails, chat, etc. Certainly, the written word can lead to confusion as can the spoken word. We've all encountered failed attempts to use sarcasm in an email or text message. For those whose command of English isn't quite the same as their American counterparts, written communication can be easier. Often, English as a Second Language (ESL) folks tend to be better communicators in writing than when speaking. Written communication also offers more time for deciphering the meaning, thinking about the message, and composing an appropriate response, whereas doing the same on a live call might create an awkward pause.

This isn't important just for customer-facing roles. It's equally relevant for internal discussions so you avoid siloing overseas staff and leaving them functionally disconnected from their colleagues elsewhere. Overseas offices experience this challenge anyway, but with a fully remote company, everyone is equal in this regard and uses the same communication tools, thus lowering the burden for those whose English may require more effort.

In short, already being a fully-remote company makes global expansion an easier, less risky, more measured step-by-step process. If your business plan calls for global offices in the foreseeable future, expanding remote work at home is a great way to start climbing the learning curve in advance.

CHAPTER 9

LESS TIME WITH LAWYERS
Risk Mitigation of Office Mishaps

As every executive and most managers know, too many business decisions are driven by a fear of litigation. What do we need to do to avoid being sued or investigated for X, or if we do, to guarantee that we can survive? For the record, let me state: I am not a lawyer, and nothing discussed here should be constituted as legal advice! Many business scenarios necessitate dealing with a well-qualified attorney.

That said, you can certainly minimize clearly foreseeable legal conflicts. The problem is that many possible approaches interfere with doing business. For example, if you don't have any employees, you'll avoid a whole host of potential liability, but you'll also be limited as to scalability and control. Some business enterprises are based around walking this tightrope, most notably "gig work" companies like Uber.

Exploring the boundaries between traditional employment and using contractors is not really the point of this discussion, though. The assumption is that most of your staff, at least in your home country, utilizes a traditional employment model. Given the necessary compliance with local rules, a fully-remote workplace by its nature reduces the risk of unfortunate and costly litigation.

—— Location Liabilities ——

The concept of "slip-and-fall lawyers" is a stereotype for good reason: businesses traditionally must carry multiple levels of liability and other insurances because of potential lawsuits. A surprising number of these potential problems are related to physical presence and location, as the trope implies: think about a customer slipping on the sidewalk and breaking their leg, or an employee crushing their fingers in a file cabinet.

Consider how remote work inherently reduces these issues. Of course, there won't be customers walking into someone's remote office, thus no possibility of an accident - nobody ever broke their leg by making a Zoom call! While there may be some workers-comp claim potential even with remote work, moving activity online reduces or eliminates a great deal. Most remote organizations have all-electronic filing systems, not physical file cabinets that can tip over and crush somebody.

Even risks that apply to home offices and regular offices equally may be reallocated in a way that benefits the company without harming anyone. If your office catches fire and burns down, you have a major business continuity challenge. If your employee's home burns down, and with it their home office, hopefully you'll provide them with sympathy and support - but it won't cripple your business, and probably won't even lose any company data. Should there be an injury in an office fire, you're probably looking at a lawsuit, but there's a much lower chance of that if the office is at home.

—— Harassment ——

Certain categories of business risk are inherently associated with physical proximity. Sexual harassment complaints frequently include physical aspects, whether it be standing uncomfortably close, unwanted touching, or cornering in the hallway or office. With a virtual workspace[20] this physical opportunity is limited, since each worker is alone in the privacy of their own home office[21].

Does this mean harassment is impossible? Of course not. Any environment provides ample room for inappropriate verbal statements or written words, but in these cases, the worst possibilities of a "he said, she said" unwinnable argument can be avoided. Did

---

[20] Carrazana, Chabeli. "Remote work may help decrease sexual assault and harassment, poll finds." 19th Polling. https://19thnews.org/2023/09/poll-remote-work-decrease-sexual-harassment/ Accessed Jun 30, 2025.
[21] Tsipursky, Gleb. "Back to the office, back to getting harassed?." The Hill. https://thehill.com/opinion/technology/4754300-sexual-harassment-in-office-return/mlite/ Accessed Jun 30, 2025.

someone send an inappropriate email? The email exists as hard evidence that can be examined and adjudicated as necessary, without relying on fallible memories or biased testimony. Similarly, inappropriate statements made via chat can be screenshotted and, in some systems, reviewed in administrator logs. Again, the recorded facts of what occurred are beyond dispute and can be addressed on their own merits.

What about statements made in a private videoconference? The first time that happens, yes, you may not have any evidence preserved. For damaging litigation to occur, however, a pattern of harassment generally must be proven. This means the behavior has to take place more than once, and as the saying goes, forewarned is forearmed. If the victim anticipates an online meeting with the harasser, all major videoconferencing systems have the capability to record that meeting. The end result will once again provide evidence to help quickly and fairly resolve complaints.

Human beings are what they are, and no matter what efforts we make, some employees will act inappropriately. The nature of remote work cannot fully resolve human nature, but the separateness and ability to record interactions can help to minimize the risks.

—— For-Cause ——

Many American jurisdictions consider employment to be "at will," that is, an employer can dismiss any employee at any time without providing any justi-fication. Terminated employees do, however, have increasing legal avenues to sue for wrongful dis-missal, as well as regulations limiting the scope of "at will" employment. As a result, corporate best practice has evolved to recommend, wherever possible, thor-ough documentation of an employee's failure to meet requirements or conform to regulations.

In a well-regulated remote work environment, you employ a whole host of metrics, records, proofs, and other documentation of the work each employee has or has not been doing. Their work computer will show activity, as will their work calendar. Their work prod-uct and its changes over time will likely be recorded in the company cloud. Most cloud storage systems don't just store documents in their current version, but also record the timing, nature, and source of changes. If an employee is terminated for lack of productivity, the proof is readily available. If it isn't, then that raises questions about management effectiveness, and orga-nizational changes must be made which we've dis-cussed earlier. All these tools are equally available for in-office work, of course – but they aren't an absolute management necessity the way they are for remote environments.

Many experts in employment best practice recom-mend the documentary use of a Performance Improve-ment Plan, in which the employee's supervisor records how the employee isn't measuring up as well as what they need to do to achieve the required standards and presumably remain employed. Everything needed for this should already be available to the manager; otherwise, how can they be expected to be properly managing employees wherever they're located?

Unfortunately, I have had to activate this process occasionally. In every case, the employee did not dispute the termination since the evidence was clear and convincing. Yes, some people are argumentative by nature, but the more solid your evidence, the less likely they'll prosecute a case against the company. Using remote work best practices makes this far easier.

Of course, a company never wants to have to terminate an employee for poor performance. This may mean they have not been fully earning their paycheck for some period of time, and that the time and effort invested in training them is at risk of being lost. The best possible outcome for a PIP is that the employee will effectively address their weaknesses and move forward as a productive employee.

Again, a well-regulated remote work environment makes this easier because the appropriate metrics are present, clear, and readily available. If the problem is timekeeping, access can be granted to your recording mechanism of choice so the employee can see how they're doing and change their habits. If the concern involves productivity, they should be able to view their score along with their manager so they can respond with improvement.

Another benefit to this sort of visibility is that while terminating people is traumatic for the person being terminated, it's often not easy on the manager either. The best managers care about their team as human beings and don't like to see them in distress. Being able to look in the mirror and know that they did everything within their power to remedy a situation before termination becomes necessary helps managers sleep better at night. It also makes them more willing to take necessary action rather than put up with an unhealthy and corrosive situation because required action is simply too painful.

—— Strategic Basing ——

One of the geniuses of America's founders was identifying the concept of the United States as a union of "free and independent States." Yes, our federal government operates with regulations, but we also have 50 states, additional territories, and a multitude of smaller jurisdictions with a surprising breadth of power to affect business operations within their sphere.

In a fully remote company, much of this is irrelevant. If everyone is working out of their own home, you don't have to worry about zoning ordinances, parking requirements, pollution control, or a host of other regulations. At a minimum, you will need to register in each state where you have an employee as a foreign corporation, pay for worker's compensation insurance, and remit appropriate tax withholding. This means that you've created a legal "nexus" since your company exists in that state. Having "nexus" requires you to obey local laws like worker's comp and other employment regulations that are handled differently state by state.

The legally-required existence of "nexus" does however provide potential benefits. If you are a fully-remote company with employees in multiple states, you likely need to purchase a national health plan so all your employees can use it no matter where they live. A number of large insurance companies like Aetna, UHC, and Blue Cross/Blue Shield are represented by insurance brokers who can help you choose the best plan.

Even national plans, though, are purchased "from" some particular state or office, and you may find an enormous difference in cost between states. You'd think this would be affected by exactly which states your employees live in, and how many of them are in which ones, but not as much as you'd expect. Even allowing for this, the official state of purchase can provide a savings of as much as 25%.

Since you're legally registered in every state where you have an employee, and therefore have a "nexus" there, often that qualifies you to buy insurance within that state. A good insurance broker can quote the exact same insurance from the different states in which you operate to find the cheapest quote. You don't have to purchase insurance in the state where your company was created, otherwise most big companies could only buy health insurance via Delaware since that's where many of them have their original registrations!

Make yourself aware of corporate tax advantages to incorporating in certain states. As with the well-known example of Delaware, you don't need any executives or even employees in a state to create a corporation there, just a legal representative. Every state has companies which provide registered agent services for this purpose, including national services like Northwest Registered Agent and ZenBusiness that can provide you with a legal presence almost anywhere, often for under $100.

CHAPTER 10

WHAT, NEVER? WELL, HARDLY EVER

When Zoom Alone Isn't Enough

Let's be realistic: for most professionals, the idea of a work environment where you never meet your colleagues in person seems odd. While many people appreciate the flexibility and comfort of no commute, only ever seeing people on their computer screens can be off-putting. Is it really true, they wonder, that a successful business can be completely run that way all the time?

Yes, it can – in certain cases. I know people I've worked with for decades who, to this day, have never physically met. For most people and for many different jobs, the difference between never meeting anyone and *occasionally* meeting them, or even just once, is vast. For this reason, the idea of a "fully remote company" shouldn't be a doctrinal statement. Yes, most work will be done out of individual employees' home offices, but whenever an opportunity presents itself, encourage physical meetups.

—— Serendipity for the Win ——

As we've seen, one of the great advantages of a fully-remote company is that you can hire the very best people from anywhere they happen to live. That said, you'll sometimes hire "clusters" of employees who don't live too far from each other. Similarly, many people do like to travel, and they may find themselves in another city for non-work reasons where a colleague happens to live.

It's easy, affordable, and legal to create a policy that employees can enjoy a meal together at company expense, regardless of any official company travel. Even the most expensive steak dinner is a small fraction of the cost of a fully paid company trip including travel and hotel accommodations, and it helps build informal networks and camaraderie within the company. Inevitably, employees will discuss business over dinner, which makes it tax-deductible. More importantly, the employees will become better acquainted as people. Commonsense limitations – like no more than once a month with the same colleague – can ensure reasonable bounds. Considering the vast sums you're saving by not paying for physical office space, this is a valuable investment in strengthening company culture.

If you happen to have a local cluster of employees, particularly if they're on the same work team, consider organizing a semi-official quarterly gathering. Once others in the company learn that this is encouraged, you'll likely find similar events sprouting company-wide, simply by taking advantage of each person's own choice of living place and personal travel.

—— Industry Gatherings ——

Many industries organize trade shows, and your partners, customers, and even competitors expect your company to attend. The trade show space has changed significantly since COVID, with many closing down or going virtual. The cost of attending, much less sponsoring an exhibit booth, has grown enormously over the same time, so many companies have reduced that budget line item.

Whether it's a wise investment in your particular case involves many factors. For a remote company, however, an industry show provides a fixed spot on the calendar where it may make sense to gather a group of your employees. The show may even be the perfect occasion for people from various departments to meet together outside of the trade floor and official events. At OwnerRez, an industry show was often the first time employees met colleagues in person, adding more value to the investment overall.

———— Company Retreats ————

It's become trendy in some circles to schedule lavish offsite corporate retreats at a resort or other grand location, often with a well-known thought leader to preside over workshops and inspirational lectures. With a fully-remote company, another factor is involved. As we've seen, top productivity doesn't necessarily require physically working together all the time, but most people feel closer to those with whom they've actually broken bread. A corporate retreat offers additional benefits that wouldn't apply where everyone sees each other every workday anyway.

Corporate retreats sometimes have a reputation for bad behavior like intoxication and the types of in-person sexual harassment discussed earlier. If you've done your job of thoroughly vetting new hires, hopefully you will have minimized any likelihood of problematic interactions.

No matter what type of retreat you're considering, the cost is certainly significant, especially for a fully-remote company because your team is more far-flung and will incur higher travel expenses. You can't just reserve a chartered bus. For any retreat, even the most lavish, the savings of no office space exceed the infrequent expenses of lodging, travel, and entertainment.

At OwnerRez, my biggest concern was how to provide support for ongoing operations during the retreat. Since the primary objective is for the entire company team to meet each other, that's more of a challenge than the traditional executive retreat or "top perform-ers" award event. We planned carefully well in advance to pull this off! Somewhat to my surprise, the benefits more than justified the bother. Actually meeting each other in person, even if only once, does improve team cohesiveness and bolsters idea generation.

As far as logistics, rarely can *every* last employee attend, whether because of family obligations or last-minute illness. Those who stay home may be able to take up the slack for essential operations like support. Otherwise, since the company is paying for the trip, you can still schedule a few hours of produc-tive work time each day, enough to keep the lights on. Since everyone is used to working remotely, they sim-ply bring their laptops with them. Nearly every resort provides adequate high-speed Internet service and likely a dedicated conference room your team can use, although you may want to send a lucky advance team to confirm this beforehand!

At our first company retreat, we engaged a noted industry thought leader, Matt Landau. While he talked about various industry topics, his emphasis was more on thinking through your own personal goals and approaches. As an executive, I found this valuable in shedding additional light on staff motivations and struggles, with an eye towards helping every member of the team achieve what was most important to them, as well as to the company. Could this exercise have been conducted remotely? I suppose so, but I think not as effectively, and certainly not as visibly. What's more, the very nature of a rare event adds to its psychological impact. It's not like "just another meeting" in the corporate conference room.

One unusual potential benefit is, in a way, very old-school. OwnerRez allowed each employee to bring a plus-one, often their spouse. Since half of the participants weren't employees but were somewhat familiar with the company, we benefited from their outsider's point of view. The heavy presence of non-employees helped keep the retreat light in tone rather than simply devolving into a hard-nosed work meeting. The plus-ones made sure the scheduled work events and working time didn't run over by too much! In the past, companies would often sponsor a Christmas party or summer picnic and invite families, encouraging social interactions outside of work. These seem to have gone out of style in recent years and aren't practical for a remote company, but a company retreat can provide many of the same benefits and strengthen camaraderie.

CHAPTER 11

ROADS NOT TAKEN

Looking at Remote Work as an Innovative Opportunity

This book has mostly been a discussion of the advantages of establishing a work-from-home operation and advice for doing so. Even at the height of COVID, though, only around half of American workers did, in fact, work from home. Since the end of the pandemic, that number has dropped to perhaps a quarter – still significant, but nothing close to a majority. And it won't ever be *everyone*, at least not in our lifetimes. Many jobs simply cannot be performed from the comfort of your bedroom. Manufacturing, construction, public safety, medical services: while a few corner-case roles can operate remotely, those industries may never go fully remote.

Some related jobs that could technically be done remotely can still benefit from physical co-location. While production engineers these days design in front of a computer which could just as well be at their home, they benefit from walking onto the factory floor and observing their designs in operation.

I am no techno-evangelist, pontificating that my preferred flavor-of-the-week is the One True Way to Do Everything that will inevitably rule the world. In fact, I'll come right out and say it: it won't. Remote work is not for every person, every company, every industry, or any "every" at all. It will, however, change the world and the way we do business in it. As a leader or employee, remote work may indeed not be for you or your organization, just like any other business practice or methodology may or may not fit. Your job is to carefully consider the pros and cons and to come to a decision whose *why* can be thoughtfully explained to the stakeholders. You'll also be better prepared to make adjustments as additional data is gathered and new tools become available that alter the move's calculus.

Business history contains countless examples of businesses that didn't innovate until it was too late for them – think of the classic example of buggy-whip manufacturers. Far worse are the companies that *did* innovate, coming up with "the next big thing" – then left it on the shelf for someone else to pick up and crush them with. Kodak had a hundred years of experience making camera film, and invented the digital camera themselves – but chose not to sell it, fearing it would destroy their film sales. Yes indeed, digital cameras did in fact destroy film sales, and Kodak with them.

Every company is now in a similar situation: thanks to COVID, we all have experience operating in a remote environment. It may or may not have been a *good* experience, but the opportunity to learn was and is present. What did you learn from those days? Or were you simply counting the moments to "get back to normal"? If the latter, ask yourself: what *other* new innovation was simply a matter of waiting it out until it blew over? Online shopping? Social media? Smartphones? Of course not: all those changed the world permanently, and remote work will do the same — though not necessarily in the same way for every single company, or for every business leader.

If you are leading a small business poised for growth, this is the perfect time to envision what a future larger version of your business looks like. If you don't already have a fixed office space, congratulations! Hopefully this book has given you ideas, tools, and guidance to help you build your organization without that infrastructural overhead. If you do have an office, you still have the opportunity to evaluate which potential jobs and roles need to be co-located and which can be performed remotely. Maybe your work requires special tools that must be on a factory floor, so those workers can't be remote, but the accounting staff can enter invoices and payroll from anywhere, reducing those unnecessary office costs.

If your business already has established itself with physical locations, your job is more challenging. Converting a traditional workspace to fully remote presents many problems, as we all learned during COVID. Some people simply can't make the transition effectively. For others, a change leads to lost efficiency and productivity. Still, certain departments can benefit from switching to remote, and new teams can be established as remote from the outset.

If you work for a large corporation that required remote working during COVID but now is trying to return staff to the office, you are now armed with rational arguments against this policy. Let's be honest: that edict probably comes from "On High," and you may not be able to persuade them to change their mind. In that case, you may need to carefully consider the likely future of that company, your place in it, where you are in your career, and the tradeoffs and risks inherent to life. Sometimes unexpected problems like the pandemic throw people out of their accustomed groove and open their eyes to better opportunities elsewhere.

And if you are a business leader who is strongly opposed to remote work, it's time to ask: why? There are legitimate reasons why remote work may not be appropriate for specific businesses, but force of habit is not one of them. Do you not trust your team? (Are you incapable of hiring trustworthy employees?) Are your subordinates too challenged by technology to make it work at home? (Or maybe it's just you?) Make your arguments, apply them to your specific business - and then see how persuasive your stakeholders like investors, business peers, subordinates, and line employees find them to be. You may be right - or you may learn something valuable, even business-critical.

I'll close with a quote from the English Enlightenment poet Alexander Pope:

Be not the first by whom the new are tried,
Nor yet the last to lay the old aside.

ACKNOWLEDGMENTS

Paul and Chris, without whom OwnerRez would never have existed.

My team at OwnerRez, who made everything possible and taught me so much. They are the true authors of the lessons covered in this book. In particular, Lydia and Adria, whose many hours of working together and talking things through helped bring clarity of focus when it was most needed.

The crew at Ripples Media, who shed light on my thoughts from different angles, helping to bring you a better-rounded and more accessible sharing of the lessons I've learned.

My father, who mostly worked from home before it was even recognized as an option, and long before today's tools – just a phone, a fax machine, and a pre-Internet computer. From these pioneering efforts, we now have so many more comfortable and effective capabilities.

Most of all, my wife and kids, for their encouragement and patience as this book slowly took form. And they lived through it themselves, from a different perspective!

ABOUT THE AUTHOR

Ken Taylor has spent his life in and around tech start-ups, beginning with a ringside seat to his father's own ventures and continuing through many more. His experience in building world-class teams and efficient processes across a wide range of industries at all stages of the business cycle, gives him a unique perspective on the needs of all business stakeholders – owners, investors, executives, managers, line workers, and customers or clients.

While Ken has had a great deal of experience of home-based work, his first experience at a *fully* remote organization came at OwnerRez, where he was the organization's first non-founder employee. At the time, the company was small enough that everyone worked out of their home offices, and by the time the company grew to a larger size to consider physical office space, COVID hit. The company never looked back, and the lessons executives learned during that time form the foundation of this book.

The ultimate goal of excellence requires constant learning and adapting, as this book explores. The pursuit of excellence also requires an understanding of history and the wisdom of the past, which led to his previous co-authored book, *The Confucian Cycle: China's Sage and America's Decline.*

Ken lives near San Antonio, Texas, with his wife Angela, their dog Magenta, and an ever-changing number of cats who often appear on conference calls made from his home office. He is the father to Veronica, who lives in Amsterdam with her husband, and Henry, who's finishing up an engineering degree. Right now, Ken is working with several startups and is working on his next business book, addressing systemic failures and proposing more effective solutions in modern hiring practices.

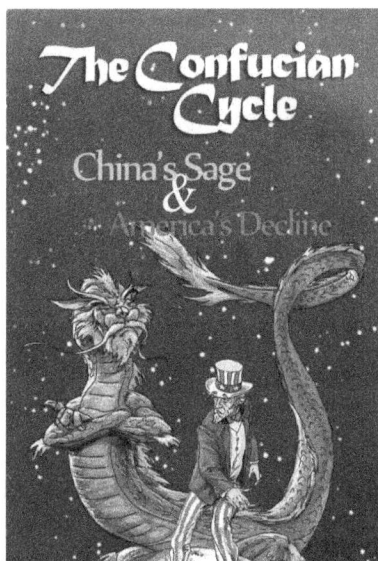

The Confucian Cycle
China's Sage
&
America's Decline

OTHER TITLES

The Confucian Cycle: China's Sage and America's Decline

The Chinese sage Confucius observed 2,500 years ago that all governments follow a cycle: from unity, through prosperity to stagnation, then to collapse and anarchy.

He taught that when government officials sought personal power or wealth instead of serving the people, society lost the "Mandate of Heaven" and fell apart.

By "Mandate of Heaven," Confucius meant that God Himself had directed how society should work. Chinese history shows 15 or 20 collapses when its government lost virtue and the country broke apart in civil war, but whenever the Chinese followed Confucius' rules, Chinese society worked well.

From his day to ours, civilizations all over the world have followed the same cycle Confucius observed. Today's United States is well into the "stagnation" phase and many observers predict a collapse. However, America has an advantage Confucius never imagined. Unlike the Chinese, America's voters have the power to replace their rulers and reform their government without armed revolution.

The authors, William and Ken Taylor, have a unique view on Chinese history. William's parents were American missionaries to Japan right after the Second World War. The Japanese had learned the secrets of civilization from the Chinese, so Confucian ideas were prevalent.

Ken, meanwhile, worked for a Japanese tech company in the mid-1990s, presenting him with a unique view of modern Confucian management. At the time, most Japanese employers were very large, very old, or both. The Japanese had recovered from the destruction of World War II by learning from other countries.

The author's experiences in the Far East help them explore how Confucius and Chinese history should inform how we view modern American politics. The Taylors' wide-ranging tour culture and modern news sheds new light on how the past both predicts the future and can be leveraged to alter it for the better.

www.ingramcontent.com/pod-product-compliance
Lightning Source LLC
Chambersburg PA
CBHW031858200326
41597CB00012B/468